GLOBAL WARMING

A Reference Handbook

GLOBAL WARMING

A Reference Handbook

David E. Newton
College of Professional Studies
University of San Francisco

CONTEMPORARY WORLD ISSUES

ABC-CLIO

Santa Barbara, California
Denver, Colorado
Oxford, England

Library of Congress Cataloging-in-Publication Data

Newton, David E.
  Global warming : a reference handbook / David E. Newton
    p.   cm.—(Contemporary world issues)
  Includes bibliographical references and index.
  1. Global warming—Handbooks, manuals, etc.
  I. Title.   II. Series.
  QC981.8.G56N48 1993   363.73'87—dc20   93-24821

ISBN 0-87436-711-5 (alk.paper)

99 98 97 96 95 94   10 9 8 7 6 5 4 3 2

ABC-CLIO, Inc.
130 Cremona Drive, P.O. Box 1911
Santa Barbara, California 93116-1911

The book is printed on acid-free paper ♻ .
Manufactured in the United States of America

For Jim Greenhaw:
Some people are special because they compose great symphonies, write great sonnets, or create great paintings. Others are special just because of who they are.
Thanks for being who you are.

Contents

List of Tables and Figures

Preface

THE WORLD SEEMS TO BE CONFRONTED with a new "Impending Environmental Disaster" every few years. Over the past two or three decades, impending environmental disasters have included nuclear holocaust accompanied by nuclear winter, devastation caused by acid rain, depletion of the ozone layer, massive famines in various parts of Africa, decimation of the world's tropical rain forests, and contamination of vast parts of the oceans by oil spills.

In 1988, yet another impending environmental disaster appeared on the horizon: global warming. Aroused to some extent by the unusually hot summer of that year, scientists, government officials, journalists, and ordinary citizens suddenly became aware of a potential global catastrophe, a change in the Earth's climate that could rival anything seen in thousands of years.

The culprit behind this scenario was a flood of carbon dioxide and other gases released by human activities. Many scientists feared that the addition of these gases to those already present in the atmosphere might drastically increase an already well-known warming phenomenon, the greenhouse effect. They predicted an increase in sea levels that might inundate coastal cities, changes in regional climates that could significantly alter agricultural patterns around the world, and a variety of other changes in the natural world and in human societies.

As is often the case with impending environmental disasters, stories in the media often featured the most extreme predictions about global warming. Prompt and drastic action was sometimes demanded to prevent the worst consequences of climate change.

The science of global warming, however, is more complex than the average citizen was led to believe. The evidence for significant, long-term changes in the Earth's annual average temperature was not particularly strong. Computer models relating concentrations of carbon dioxide and other gases to temperature trends were not well developed.

Projections about environmental effects that might be expected as a result of temperature changes in the atmosphere were subject to serious questions.

All in all, responsible scientists differed in their opinion on whether global warming had begun and, if so, what effects it ultimately might have. Those differences of opinion have created difficult choices for government officials. The kinds of changes needed to curtail global warming are likely to be very expensive. In view of the questions surrounding global warming and in view of the weak nature of the world's economy, policy makers have wondered what actions are justified in dealing with this latest impending environmental disaster. That question is not likely to be answered soon.

The purpose of this book is to provide sufficient background information about global warming to allow concerned citizens to contribute to the decisions that will ultimately be made. The first three chapters of the book provide a historical account of the global warming debate with some biographical sketches of important figures involved in that history.

Chapter 4 offers a collection of facts and opinions about this debate. Chapters 5–7 provide resources for those who wish to pursue the study of global warming in more detail, a list of organizations interested in the issue, and print and nonprint references on the topic. A glossary of terms used in the study of global warming and its effects rounds out the book.

1

Changes in the Earth's Climate

PLANET EARTH IS CONSTANTLY CHANGING. Sometimes change is abrupt and dramatic, as during an earthquake or volcanic eruption. But in most cases, change comes about very slowly over hundreds or thousands of years, and humans are hardly aware of it.

Such is the case with climatic change. Suppose you could be transported back 200 years in time. The weather conditions you would find in the 1790s would be much the same as those you experience today. In terms of climate change, 200 years is but the blink of an eye.

Yet, many scientists are convinced that our climate is changing. They point to statistics that show the Earth very slowly has been getting warmer, especially over the past 150 years. For many people, those statistics are bad news. As the Earth becomes warmer, it is likely to undergo some important changes, they say. Ocean levels will rise. Coastal cities will be flooded. Productive farmlands will become deserts. These threats are serious enough, these people warn, that actions will have to be taken soon to prevent widespread disaster.

Other people disagree. Some do not think data on weather patterns are clear. We just do not have enough information, they say, to predict long-term trends. Others agree that a warming trend has occurred, but they disagree about the cause of the trend, its possible effects, or what to do about it.

The problem of global warming presents a profound challenge to the world. Nations must make decisions on an issue about which serious disagreements exist among scientists. If we wait too long to take action, it may be too late to stop the warming trend. Horrible disasters could result. But if we move too quickly and there turns out to be no real problem, huge amounts of money could be wasted.

Weather and Climate

The term *weather* refers to conditions in the Earth's atmosphere over time, such as a day or a week. Is it raining? How hot is it? From what direction and how fast is the wind blowing? Factors such as these are the weather conditions.

Climate refers to weather patterns over a longer time, usually no less than 30 years. The weather this summer might be unusually hot. Or the region you live in may have had a drought that lasted for six years. But neither of these events tells us anything about climate change. Weather patterns change from year to year and decade to decade, and these fluctuations are normal. But they do not necessarily indicate long-term climatic trends. Only when consistent patterns can be traced over many decades is it possible to say that the climate has begun to change.

All of the factors that make up weather and climate—precipitation, temperature change, air movements—take place in the Earth's atmosphere. To understand weather and climate, therefore, it is first necessary to learn about the Earth's atmosphere.

Earth's Evolving Atmosphere

Primitive Earth's atmosphere was very different from what it is today. It was, in fact, more like the atmospheres found on Saturn, Jupiter, Uranus, and Neptune. Methane, ammonia, hydrogen, carbon dioxide, and carbon monoxide were probably the most common gases in the primitive atmosphere. Virtually no modern-day life form could survive in such an atmosphere.

At some point, however, changes began to occur. Simple life forms began to appear. These life forms had the ability to metabolize carbon dioxide, water, and other substances, releasing oxygen gas in the process. As green plants began to appear, the conversion of carbon dioxide to oxygen by photosynthesis took place more rapidly. Over time, the concentration of oxygen in the atmosphere increased, and the amounts of methane, ammonia, carbon dioxide, and other primitive atmosphere gases were reduced.

The percentage of oxygen in the atmosphere rose from zero at the time of Earth's creation, to about 0.4 percent 800 million years ago, to about 4 percent 580 million years ago, to its present level, 21 percent, about 50 million years ago. Today, the concentration of primitive atmosphere gases is very low. Methane makes up only 1.5×10^{-4} (15/100,000) percent of the atmosphere; ammonia, 1×10^{-6} (1/1,000,000) percent; hydrogen, 5×10^{-5} (5/100,000) percent; carbon monoxide, 1.2×10^{-5} (12/1,000,000) percent; and carbon dioxide, about 0.0353 (353/10,000) percent.

In some ways, the most important relationship among atmospheric gases involves oxygen and carbon dioxide. The consumption of one of these gases usually results in the production of the other. For example, when plants grow, they take carbon dioxide out of the air and use it to produce starch, cellulose, and other carbohydrates—the process called photosynthesis. Oxygen gas is released as a by-product of this reaction:

$$\text{carbon dioxide} \xrightarrow{\text{is converted to}} \text{oxygen}$$

When plants die and decay, this process is reversed. Plant material reacts with oxygen in the air (it "oxidizes") to produce carbon dioxide. Combustion (burning) and respiration by animals involves a similar process:

$$\text{oxygen} \xrightarrow{\text{is converted to}} \text{carbon dioxide}$$

The reactions involving oxygen and carbon dioxide reached equilibrium in the atmosphere many millions of years ago. The term *equilibrium* means that the rate at which oxygen and carbon dioxide are used up in these reactions is the same as the rate at which they are being produced. The concentration of each gas has remained nearly constant, therefore, for millions of years.

The Structure and Composition of the Atmosphere

As a matter of convenience, scientists usually divide the Earth's atmosphere into four major regions. The region closest to Earth, the *troposphere,* extends to a height of 10 to 16 kilometers (6–10 miles) above the Earth's surface. Beyond the troposphere are the *stratosphere* (16–50 kilometers or 10–30 miles), the *mesosphere* (50–85 kilometers or 30–50 miles), and the *thermosphere* (beyond 85 kilometers or 50 miles).

The part of the atmosphere of greatest interest to humans is the troposphere, because that is the region in which nearly all weather takes place. About 95 percent of all gases found in the atmosphere are found in the troposphere. Nitrogen, oxygen, and argon—the most abundant gases—make up 78 percent, 21 percent, and 0.9 percent, respectively, of the troposphere. Carbon dioxide, neon, helium, methane, and krypton are the next most common gases in the troposphere.

Tropospheric gases have a number of important functions. As mentioned earlier, for example, plants need carbon dioxide and animals need oxygen to survive. But gases in the atmosphere also play critical roles in the way the Earth gains and loses energy.

By far the most important source of energy for the Earth is the sun. Each minute, 1.12×10^{19} calories of solar energy reach the Earth's atmosphere. That energy arrives in the form of X rays, gamma rays, visible light, ultraviolet radiation, infrared radiation, and other forms of solar radiation.

Solar radiation reaching the Earth's atmosphere experiences a number of different fates. About 40 percent of the radiation is reflected back into outer space by dust and clouds in the atmosphere. Ten to twenty percent of solar radiation reaching the atmosphere is absorbed by gas molecules in the atmosphere. For example, a form of oxygen known as *ozone* captures ultraviolet light in the stratosphere. Most gases capture at least some form of solar radiation. This capture of solar radiation by atmospheric gases increases the amount of heat stored in the atmosphere.

The remaining 40 to 50 percent of solar energy passes through the atmosphere and reaches the Earth's surface. About one-third of this radiation is reflected back into the atmosphere off snow, water, rocks, soil, and other materials on the Earth's surface. The other two-thirds are absorbed by these materials. This absorbed solar energy is given off as heat waves (infrared radiation) back to the atmosphere.

Much of the infrared radiation returned to the atmosphere escapes back into space, but not all of it. Carbon dioxide, water, and certain other gases in the troposphere capture and hold some of the radiation. These gases retain some of the solar energy that otherwise would have escaped back into outer space, raising the temperature of the atmosphere.

Carbon dioxide and other tropospheric gases act like the glass windows in a greenhouse. Visible light is able to pass through the glass windows on its way into the greenhouse. When objects inside the greenhouse absorb visible light, they become warm and reradiate heat (infrared radiation). Because infrared radiation cannot pass through glass, however, the greenhouse becomes warmer. Due to its similarity to heating in a greenhouse, the capture of infrared radiation by tropospheric gases is sometimes called the *greenhouse effect.* Atmospheric gases such as carbon dioxide that capture heat are sometimes referred to as *greenhouse gases.*

The greenhouse effect is critical to the survival of life as we know it on Earth. Without the capture of heat by carbon dioxide and other greenhouse gases, the Earth's surface temperature would be about 35° C (63° F) lower. At that temperature, water would not exist in the liquid state. Lakes, ponds, rivers, streams, and most of the oceans would be frozen solid.

The amount of solar radiation reaching the Earth's atmosphere changes over time. Scientists have long known, for example, that the number of sunspots on the sun increases and decreases in an 11-year cycle. There is some evidence that weather patterns on Earth tend to follow that cycle. No explanation has been found for this pattern, however.

In the longer term, the amount of solar radiation received by the Earth changes because of natural variations in the Earth's orbit around the sun. For example, the planet's angular orientation toward the sun varies between 22° and 24.5° every 41,000 years. This variation means the Earth will receive greater or lesser amounts of solar energy during the 41,000-year cycle. The ice ages and other glacial periods appear to be the result of changes such as these, rather than major changes in the Earth's atmosphere.

The Study of Ancient Climates

Scientists know that the Earth's climate has changed significantly during the planet's history. At one time, the world was warm enough to permit the growth of abundant vegetation as far north as the island of Greenland. In fact, the island got its name because of the lush forests that once grew there. Now, of course, Greenland is almost totally covered with thick layers of snow and ice.

The best-known climatic changes are the ice ages and the interglacial periods that occurred between them. The Earth apparently has experienced seven major ice ages in its history. Each ice age, in turn, has consisted of alternate periods of warming and cooling. The most recent ice age began about 125,000 years ago and has been marked by five major cycles of cooling and warming.

Attempts to understand the ice ages go back at least 100 years. Over time, it has become apparent to scientists that a combination of astronomical factors account for most of the long-term changes in the Earth's climate. These factors include changes in the planet's angular tilt, its axial precession (a "wobble" that occurs in the planet's orientation to the sun on a 23,000-year cycle), and a 100,000-year variation in the shape of the planet's orbit around the sun. The Serbian engineer Milutin Milankovitch worked out a mathematical theory that shows how these factors interact to produce major changes in the Earth's climate.

The effort to explain climatic variations continues today. Scientists operate on the assumption that the more they know about natural climate change, the better they will be able to understand changes brought about by human activities. Today, researchers realize that, in addition to astronomical factors, a complex interaction of conditions on the planet affects climate change. The concentration of atmospheric gases, cloud cover, and the oceans are probably the most important of these conditions. An example of current research in this area is the work of Columbia University geologist Wallace Broecker.

Broecker has studied a period of Earth history known as the Younger Dryas. The Younger Dryas began about 11,000 years ago, at a time when the Earth was gradually warming, recovering from the most recent ice age. In less than a century, the warming pattern changed dramatically in one part of the planet, the region adjacent to the North Atlantic. In that region, temperatures dropped as much as 6° C (11° F) in about 50 years. Canada, Iceland, Great Britain, and Northern Europe became much colder, and ice formed in the North Atlantic, apparently disrupting normal circulation patterns in the Atlantic Ocean. During the 1980s, Broecker developed a theory that explains how a change in the flow of glacial meltwater from the Arctic ice sheet caused the Younger Dryas. In so doing, he showed how the movement of sea currents and climate are closely related (Lippsett n.d.).

Broecker was interested not only in learning more about the Younger Dryas but also in finding out what that period had to teach us about current climate change. He concluded that an increased release of carbon dioxide as a result of human activities could lead to a modern Younger Dryas. "If we look to the natural system for advice," he says, "the natural system says that it often reacts violently when pushed. Therefore, we shouldn't be too confident that it may not give us a few surprises if we nudge it with greenhouse gases" (Lippsett n.d.). Like Broecker, many other scientists believe the Earth's ice ages hold an important key to knowing more about future climate changes resulting from the added effect of human activities.

The Effects of Human Activities on the Greenhouse Effect

Human activities have always affected the composition of the atmosphere, and they continue to do so today. When the earliest cave people built campfires, for example, they released carbon dioxide, carbon monoxide, soot, and other gases into the atmosphere. But for most of human history, these effects were limited to small, specific regions of the planet. The overall global effects were largely insignificant. The amount of gases added by human activities was very small compared to the total volume of the atmosphere.

But that situation began to change in the late 1700s. Humans found that the combustion of *fossil fuels*—coal, oil, and natural gas—provided a very efficient source of energy. Fossil fuels began to be used to run all types of industrial machinery and, later, to operate trains, boats, cars, buses, and other forms of transportation. So dependent have humans become on coal, oil, and natural gas that some historians refer to the age in which we live as the Fossil Fuel Age.

The changes that have occurred in the last 200 years are reflected in the amounts of fossil fuels produced and consumed. Worldwide production of coal increased from 150 million tons in 1860 to 4.573 billion tons in 1987. (The unit *ton* in the British system of measurement is nearly identical to the unit *metric ton* in the metric system.) See Chapter 4, tables 4.1 and 4.2, for more information about fossil fuels production. The rate of coal production doubled every 20 years between 1860 and 1920. In the United States, coal production mushroomed from 10 million tons in 1850 to 1 billion tons in 1990. Similar patterns exist for the production of oil and natural gas.

The major components of all fossil fuels are carbon and hydrocarbons. Hydrocarbons (HC) are compounds that consist of hydrogen and carbon. Methane, CH_4, is the simplest hydrocarbon. When any fossil fuel is burned, water, carbon dioxide, and carbon monoxide are formed:

$$C \quad + \quad O_2 \quad \xrightarrow{\text{heat}} \quad CO_2 \quad + \quad CO$$
$$\text{carbon} \qquad \text{oxygen} \qquad\qquad \text{carbon} \qquad \text{carbon}$$
$$\text{dioxide} \qquad \text{monoxide}$$

$$HC \quad + \quad O_2 \quad \xrightarrow{\text{heat}} \quad H_2O \ + \ CO_2 \ + \ CO$$
$$\text{hydrocarbons} \quad \text{oxygen} \qquad \text{water} \quad \text{carbon} \quad \text{carbon}$$
$$\text{dioxide} \quad \text{monoxide}$$

In addition, small amounts of other gases are released during the combustion of fossil fuels. For example, coal, oil, and natural gas often contain small amounts of nitrogen and sulfur. When these elements burn, they are converted to nitrogen and sulfur oxides.

Of all the products of combustion, carbon dioxide has the most important effect on global temperatures. Water vapor is recycled so rapidly (through precipitation and evaporation) that human activities have almost no effect on atmospheric concentrations of water vapor. And carbon monoxide and oxides of nitrogen and sulfur have little effect on the absorption of heat in the atmosphere.

Before the Industrial Revolution, the amount of fossil fuel burned was relatively small. The Earth's atmosphere easily absorbed the carbon dioxide released during combustion. Carbon dioxide from fossil fuel combustion was used by plants for photosynthesis, for example, just as was carbon dioxide from other sources. Natural processes in the atmosphere maintained the level of carbon dioxide and other greenhouse gases at a relatively constant level. After the Industrial Revolution, however, humans began pouring carbon dioxide and other greenhouse gases into the atmosphere much more rapidly. Natural processes such as photosynthesis by plants could not remove carbon dioxide as quickly as it was being added to the atmosphere.

Slowly the concentration of carbon dioxide in the troposphere began to increase. In just the 35 years during which accurate records have been kept, the percentage of carbon dioxide in the atmosphere has increased from about 315 *parts per million* (ppm) to more than 350 ppm (see also figure 4.2). (The unit *parts per million* is commonly used to express the concentration of gases in the atmosphere. One ppm is equal to 0.0001 percent.)

This change can be compared with previous fluctuations in atmospheric levels of carbon dioxide detected by scientists who have developed methods for analyzing air trapped in ice in the polar ice caps. This ice has been in place for tens of thousands of years, and the trapped air tells us what the composition of the atmosphere was like back then. Analysis of this air shows that levels of carbon dioxide before the Industrial Revolution were about 280 ppm. In no case, during the thousands of years studied, did the concentration of carbon dioxide appear to change more than it has in the past 50 years (Houghton et al. 1989).

The combustion of fossil fuels is only one—although probably the most important—way in which humans affect the Earth's climate. The destruction of tropical rain forests, for example, also may have a significant effect on atmospheric gases. As more and more trees are cut down, some scientists say, the worldwide rate of photosynthesis is likely to decrease. Carbon dioxide will be removed from the atmosphere less quickly, contributing to an increase in the carbon dioxide concentration in the troposphere. According to some estimates, deforestation has contributed to 10 to 30 percent of the increase in carbon dioxide levels observed over the past four decades (Edgerton 1991).

But few data exist on this question. Other scientists believe that the growth of new plant life on deforested lands may actually result in an *increased* rate of photosynthesis. It is conceivable, they say, that destruction of forests could actually result in the faster removal of carbon dioxide from the atmosphere, causing a *decrease* in carbon dioxide concentration (Salati et al. 1991).

Some experts point out that forests in temperate regions are actually increasing in size. In the United States, for example, the 1700s and 1800s were a period of massive timber cutting. But the 1900s brought a more balanced approach to lumbering and the reforestation of many lands. Simply put, no one really knows the overall global trend of tree growth or how this trend may affect carbon dioxide levels in the atmosphere.

In addition, other gases contribute to the greenhouse effect. One that is receiving increased attention is methane, the main component of natural gas. One reason for the interest in methane is the gas's ability to absorb radiation. A single methane molecule is 15 to 30 times more efficient at absorbing radiation than is a single molecule of carbon dioxide. That means that, even if the amount of methane in the atmosphere is much less than that of carbon dioxide, it may still contribute significantly to atmospheric warming.

Methane is produced by the breakdown of plant materials by anaerobic bacteria. For example, when cattle digest food, the bacteria in their stomach release large amounts of methane. Cattle belch at the rate of about twice a minute, releasing, per cow, an average of about 1 kilogram (over 2 pounds) of methane each day in the process.

Rice paddies are another source of methane. Rice stalks act like tiny tubes that provide a way for methane, released in water-logged soils, to escape into the atmosphere. Without this way of escaping, much of the methane would be decomposed by anaerobic bacteria that live in the soils, and the methane would never reach the atmosphere. Thus, as humans look for ways to expand their food supplies (more cattle and rice fields), they may also be contributing to the concentration of greenhouse gases.

Another source of methane gas is termites. Like cattle, termites release methane as they digest food. Some scientists believe that this explains another way in which deforestation contributes to climate change. As more forests are cut down, dead trees will be attacked by termites, and the amount of methane released will increase.

Again, few data exist on this question. Other scientists doubt that termites are a significant factor in any contribution that methane may make to global warming. The only fact on which everyone can agree is that the concentration of methane is increasing rapidly, more rapidly than that of any other greenhouse gas. Since 1979, methane levels in the atmosphere have been increasing at the rate of 1 percent per year. The source of that increase and its ultimate effects on climate, however, are still largely unknown.

Two other compounds of increasing concern are nitrous oxide (N_2O) and chlorofluorocarbons. Nitrous oxide is produced during the combustion of fossil fuels by the breakdown of chemical fertilizers and by bacterial action, especially in tropical soils. Although the exact mechanisms by which nitrous oxide reaches the atmosphere and its effects on climate are not well understood, its concentration is increasing rapidly. Scientists estimate that 5 million tons of nitrous oxide are added to the atmosphere annually (Patrusky 1988).

Chlorofluorocarbons (CFCs) are used for a wide variety of industrial and commercial purposes. Three of their most common applications are in plastic foam insulation, as a cleaning fluid for electronic components, and as a coolant in such appliances as refrigerators and air conditioners. CFCs are a matter of concern because they contribute to the destruction of ozone in the Earth's stratosphere. The presence of ozone in the stratosphere is important to humans because ozone absorbs ultraviolet light from solar radiation. The reduced amount of ultraviolet light is important because ultraviolet light is a major cause of cancer.

CFCs are also a factor in the greenhouse effect because they absorb a band of radiation that is not absorbed by any other gas in the atmosphere. Normally this band of radiation escapes from the atmosphere through the *atmospheric window*. That term refers to the quality in the atmosphere created by the absence of gases that have the ability to trap a particular type of radiation. As CFCs accumulate in the atmosphere, the atmospheric window begins to "close," more heat is retained, and the atmosphere's temperature increases. The rate at which some CFCs are being released into the atmosphere is greater than any other greenhouse gas, in some cases at least 10 percent greater annually. By some estimates, methane, nitrous oxide, ozone, and CFCs together contribute a little more than half of the warming potential produced by carbon dioxide itself. But if current trends continue, the warming potential contributed by these four gases will be equal to that of carbon dioxide by the year 2030 (Intergovernmental Panel 1992).

The Effects of Global Warming

The study of climate change involves a few well-known facts, about which everyone can agree, and a great many uncertainties, about which there is a great deal of debate. In the former category is the greatly increased use of fossil fuels and the increase in carbon dioxide concentration in the atmosphere. Many scientists have used these fundamental data to argue a scenario of global warming. According to that scenario, the increased levels of carbon dioxide in the atmosphere mean that the Earth's annual average temperature has begun to rise. As the Earth's temperature rises, a number of environmental changes are likely to follow. The debate over global warming tends to focus on these two points: (1) what is the connection, if any, between carbon dioxide levels and temperature changes, and (2) what are the probable effects, if any, of increased atmospheric temperatures?

The measurement of temperature changes is an exceedingly difficult problem. In the first place, accurate measurements have been available only since the late 1800s. Scientists have methods for estimating temperatures back a few hundred or thousand years, but those methods do not always produce reliable data. Second, measurements of temperature in the atmosphere, in the ocean, and over land may differ from each other significantly. It is not always clear which set of data provides the best clue to the Earth's annual average temperature.

The overall trend in the past century does seem to be generally clear, however. Between 1880 and 1990, the Earth's average temperature seems to have increased by about 0.5° C (0.9° F), or about 2 percent. Studies conducted by the National Aeronautic and Space Administration's (NASA) Air Resources Laboratory have shown that the Earth's surface temperature has increased by an average of 0.08° C (0.14° F) since 1958. These changes have not been consistent, however. Between 1880 and 1940, there was a gradual, but irregular, increase of about 0.5° C (0.9° F). Between 1940 and 1970, however, temperatures experienced a decline of about 0.2° C (0.4° F). After 1970, another increase began, ending in the "greenhouse decade" of the 1980s, a period that included 6 of the 10 warmest years ever recorded (Boden, Sepanski, and Stoss 1991).

Scientists use existing temperature data to model the future. They write mathematical equations that describe energy changes in the atmosphere. Then they revise those equations until the equations fit as closely as possible to temperature data *from the past.* When the models work as well as possible for the past, they are used to predict the future.

Models often have produced very satisfactory results in describing the climates of planets other than the Earth. In most cases, however, those planets are simpler to study because they contain no life. Models of the Earth's atmosphere are much more complicated because they have to take into account the effects of human activity, forests, aquatic organisms, and other life forms. In addition, determining the influence of water in all its forms— oceans, glaciers, and water vapor in the air—is exceedingly difficult. It is not surprising, then, that different models of the

Earth's climate often disagree with each other and with the real conditions found on Earth.

Still, models have greatly improved in recent decades. Scientists now agree that their predictions have increasing validity. One prediction on which these models tend to agree is that a doubling of greenhouse gases in the atmosphere will result in a 2° C (3.6° F) increase in global temperatures. The range of predictions varies from a low of 1.5° C (2.7° F) to a high of 4.5° C (8.1° F) per doubling of greenhouse gases (Intergovernmental Panel 1992).

What would be the effect of a 2° C (3.6° F) increase in the Earth's average temperature? Again, there is no certain way of answering that question. The best scientists can do is make predictions based on past experience. One of the most likely changes would be a rise in sea levels. Warmer temperatures will cause ocean water to expand, mountain glaciers to melt, and, perhaps, portions of the polar ice caps to melt. As a result, sea levels could rise anywhere from 0.5 to 1.5 meters (1.6 to 4.9 feet) in the next 50 to 100 years.

Higher sea levels could have at least two major effects. First, substantial portions of coastal cities such as New York City, Miami, New Orleans, Venice, Bangkok, and Rotterdam might be submerged. Some low-lying nations might lose a significant fraction of their land area to flooding. For example, half of the population of Bangladesh lives on land that is less than 5 meters (16 feet) above sea level. Many of those people could lose their homes to a rising ocean.

Some reports on global warming have envisioned severe disasters for certain parts of the world. A 1984 report from the Environmental Protection Agency (EPA), for example, warned that current trends might result in a sea-level rise of more than 3 meters (10 feet) by the year 2100 and double that amount two to five centuries later (Smith and Tirpak 1990). In such a case, Capitol Hill in Washington, D.C., would become an island; the streets of Charleston, South Carolina, would be submerged; and San Francisco Bay would spread 120 kilometers (75 miles) inland to Sacramento.

Even if coastal areas are not submerged, they may experience an incursion of sea water. Miami's water supply, for example, lies only a meter (about 3 feet) underground. An increase in sea level would cause a flood of sea water into the city's natural reservoir.

A second consequence of an increase in the Earth's average temperature might be changing regional climates. Predictions vary, but all models point to increased rainfall in some parts of the world and decreased rainfall in other parts. In such a case, productive farmlands might become deserts and arid regions might become arable. In addition, a warmer Earth would mean that crop-growing regions would shift northward. According to one estimate, cropland in Appalachia, the southeastern United States, and the southern Great Plains would decrease by 5 to 25 percent, while that in the northern Great Lakes states, the northern Great Plains, and the Pacific Northwest would increase by 5 to 17 percent (Smith and Tirpak 1990).

Predictions of a 2° C (3.6° F) temperature change may be difficult to imagine or evaluate. But historical comparisons are possible. For example, during the most recent ice age, 18,000 years ago, the Earth's temperature was about 5° C (9° F) lower than it is today. During the "Little Ice Age" that lasted from 1550 to 1850, global temperatures were only 0.4° C (0.7° F) lower than present levels. An increase of 2° C (3.6° F) has not been observed in at least a million years on Earth (Hileman 1989).

A 1990 report on global warming by the United Nations Intergovernmental Panel on Climate Change raised yet another aspect of this complex issue. Authors of the report warned about the possibility of a *positive feedback* mechanism occurring during global warming. A positive feedback mechanism is one in which the result of the mechanism serves to accelerate the mechanism itself. For example, in the case of global warming, a warmer Earth might itself bring about further warming (Intergovernmental Panel 1990).

Dr. John Woods, of Great Britain's Natural Environment Research Council, explained how this might come about. Presently about one-third of the heat trapped by greenhouse gases ends up in deep, relatively cold layers of the ocean. The concept of heat's being trapped in cold water may seem a contradiction. But the ocean waters would be much colder still if they did not contain that heat. The heat is carried by vertical ocean currents that move surface waters warmed by sunlight to deeper layers of the ocean. If the world becomes warmer, Woods says, rainfall will increase and so will ocean circulation. Under those circumstances, heat stored deep within the oceans might be carried to the surface and released to the atmosphere, further warming the planet (Woods 1984).

Points of Disagreement

Some scientists, journalists, and public officials have already announced the arrival of global warming. In June 1988, for example, Dr. James Hansen, a climate expert at NASA's Goddard Institute of Space Studies, told the U.S. Senate's energy committee that he was "99 percent certain" that the apparent warming trend that was currently being observed was, in fact, a real warming trend (U.S. Government Printing Office 1988). But other experts are more hesitant to make such bold statements. They point to a number of troubling inconsistencies in the available data on global warming and in the way those data have been used in modeling experiments.

One problem involves the release of carbon dioxide into the atmosphere. No one doubts that carbon dioxide emissions have escalated in the past century. But the level of carbon dioxide *in the atmosphere* has not increased at anywhere near the rate of emissions. Between 1958 and 1990, for example, carbon dioxide emissions more than doubled, but the concentration of carbon dioxide in the atmosphere increased by only 12 percent (compare tables 4.1 and 4.2 to figure 4.2 in Chapter 4).

Scientists are puzzled by the apparent existence of a "mystery sink" for carbon dioxide. The term *sink* refers to any final location where a material (carbon dioxide, in this case) is eventually deposited. Two major sinks for carbon dioxide on the Earth are the oceans in which carbon dioxide dissolves and green plants, which take carbon dioxide out of the air in the process of photosynthesis. It is obvious that a large portion of the carbon dioxide emitted by human activities does not stay in the atmosphere, but ends up in one or more sinks, at least some of which we don't know about. Until these "mystery sinks" can be identified, we can't be sure how increasing carbon dioxide *emissions* will affect carbon dioxide *concentrations* in the atmosphere.

A great deal of controversy also surrounds temperature data. When measuring temperature differences for the entire Earth over a 12-month period, the opportunity for error can be quite large. And, as mentioned above, temperatures taken in different places at the same time have been somewhat variable. For a long time, for example, most temperature readings used in global studies were taken in urban areas. But scientists know that urban areas tend to be warmer than nonurban areas. To what extent have global temperature readings been biased by this urban factor?

Also, ocean temperature readings tend to differ markedly from those taken on land. Geophysicist Richard Lindzen, from the Massachusetts Institute of Technology (MIT), reports that there has been no observable change in ocean temperatures since the nineteenth century (Lindzen 1990). Scientists are not sure how these results can be reconciled with temperature data for land and atmosphere over the past century.

Doubts also have been expressed about the dependability of existing temperature measuring methods. According to Thomas R. Karl, a meteorologist for the National Oceanic and Atmospheric Administration (NOAA), scientists have continually changed the techniques used for measuring global temperatures over the past four decades. These changes alone may account for reported temperature changes, Dr. Karl believes. We can't even be sure, he warns, that the 1980s really were the warmest decade in history (Boden, Sepanski, and Stoss 1991).

An additional problem was created in 1988 with the release of data collected by two U.S. weather satellites. The satellites made detailed temperature measurements of the troposphere from 1979 through 1988 and found no upward trend during that decade. These results differed from measurements made on the ground, where a modest increase was observed.

On the other hand, some data do seem to support the contention that temperatures are rising. Researchers at Ohio State University reported in 1989, for example, that ice cores taken from glaciers in Central Asia show a temperature increase of 1° to 3° C (1.8° to 5.4° F) over the past 100 years. Urban factors certainly cannot account for this change, they say (Thompson et al. 1989).

Even where global warming is found, it may be difficult to tie it to the greenhouse effect. Most models predict, for example, that global warming will be most severe in the upper latitudes and least severe near the equator. But at least one study that did find global warming produced just the opposite results. It found that temperatures in the polar and North Temperate zones showed a slight decrease from 1958 to 1987 even though the Earth as a whole was warming during that time (Boden, Sepanski, and Stoss 1991).

A second study produced another puzzling result. According to most models, oceans should warm more slowly than land masses. Because the Southern Hemisphere is covered mostly by oceans, one would expect warming to occur more slowly there than in the Northern Hemisphere. But this study showed just the opposite (Boden, Sepanski, and Stoss 1991). Still, such unexpected and unexplained results have not changed the minds of researchers who are convinced about the carbon dioxide–global warming connection. David Thomson and his colleagues at AT&T Bell Laboratories, for example, announced in 1989 that their research had produced a "99.99 percent chance that the warming and the CO_2 rise are causally related" (Begley 1989).

A major point of dispute involves the possible role of clouds in climate change. It seems obvious that an increase in global temperatures will cause an increase in the rate of evaporation from lakes and oceans. The water vapor thus formed will become part of the atmosphere and, eventually, condense into clouds. The warmer the Earth becomes, therefore, the more clouds it should develop. But clouds play an important role in determining climate, too. A certain fraction of sunlight reaching the troposphere is reflected back into space by clouds. The more clouds there are, the more sunlight will be reflected, and the cooler the Earth will become.

The effect of clouds, however, is not that simple. Some clouds are bright and tend to reflect solar radiation. Others are dark and tend to absorb radiation. High and low clouds also behave differently in terms of their response to solar radiation. In addition, clouds over tropical areas have different properties and, therefore, different effects from those over middle and higher latitudes. Because it is difficult to predict the type of clouds that might be produced by further atmospheric warming, the nature of clouds' influence on climate is unclear. There is at least the possibility, however, that cloud development will be a self-correcting mechanism that will automatically keep the Earth's temperature from rising.

Another self-correcting mechanism may be the growth of green plants. As carbon dioxide accumulates in the atmosphere, the rate of photosynthesis may increase. If this were to happen, carbon dioxide levels would increase much less rapidly than might be expected.

Evidence for a possible rise in sea levels is weak, too. Although the claim is made that the Earth's average temperature has increased over the past decade, there has been no corresponding change in sea levels. Sea levels have risen about 1 centimeter (less than a half-inch) per decade over the past century, but the *rate* at which they are rising has not changed. Given the marginal accuracy of measuring devices and the influence of natural land movements, even these data are of questionable value.

One of the most troublesome matters for scientists is that they know next to nothing about some of the most important factors in climate change. The oceans are perhaps the best example. First, the oceans are an important sink for carbon dioxide. Carbon dioxide dissolves in water, but its solubility depends on a number of factors including its concentration in the atmosphere, water temperature, and salinity of the sea water. In addition, microorganisms in the ocean metabolize carbon dioxide as they grow, making them an important factor in the amount of carbon dioxide absorbed by the oceans.

Second, the oceans are a major sink for heat in the atmosphere. Water absorbs heat and then, by virtue of ocean currents, moves that heat around the planet. But scientists know very little about the way that ocean waters circulate, temperature changes that take place, and transfer of carbon dioxide in and out of the oceans. In fact, scientists know far less about ocean dynamics than they do about atmospheric dynamics.

Yet, oceanographers have evidence that changes in the oceans were closely related to the beginnings and ends of the last five glacial periods. It is nearly certain that ocean changes also will be involved in future climate changes. Therefore, our lack of knowledge about the oceans constitutes a huge hole in our ability to understand and predict future climatic changes.

Finally, a continuing problem in predicting climate change has been the limited effectiveness of modeling systems. Currently, five major modeling programs are in existence. They are located at Oregon State University, Corvallis, Oregon; the National Center for Atmospheric Research (NCAR) in Boulder, Colorado; NOAA's Geophysical Fluid Dynamics Laboratory in Princeton, New Jersey; NASA's Goddard Institute for Space Studies in Greenbelt, Maryland; and the United Kingdom's Climate Research Unit, at the University of East Anglia, Norwich, England.

These five modeling programs tend to produce good results in predicting the climates of Mars and Venus and Earth climates from the Mesozoic Era, all of which can be checked against existing data. They also tend to agree with each other on broad predictions for global changes in the next century. However, they tend to produce widely different predictions for regions smaller than a continent. For this reason, they have only limited value in dealing with questions about climatic changes in specific regions of the

Earth. Most modelers believe they will need another 5 to 10 years to refine their programs to a point where the desired accuracy of predictions can be obtained.

Social, Political, and Economic Issues

Deciding what to do about climate change is made profoundly difficult by the many scientific uncertainties. Most experts in the field have taken one of three positions. One position is that nothing needs to be done. Even if the whole global warming argument is true, some authorities say, there is no realistic way to deal with the problem. Nations and individuals would have to be asked to make fundamental changes in their lifestyles, changes they probably would not accept. Even under the best of circumstances, drastic changes might not be sufficient. For example, a 1984 EPA report, *Can We Delay a Greenhouse Warming?*, suggested that a total ban on the use of coal (a highly unlikely event) could delay global warming by 15 years, but could not prevent it (Seidel and Keyes 1983). The only solution, some people suggest, is for humans to learn to live on a slightly warmer planet and adjust to the new problems that may develop from global warming.

A second position is that we just do not know enough about the science of climate change to take any action at all. Too many questions remain to justify investing money in solutions. The U.S. government has held this position for at least a dozen years. The administrations of both Ronald Reagan and George Bush insisted that money be spent on additional research on global warming, but not on efforts to deal with the problem (if the problem even exists).

Interested officials in the U.S. government took a major step in this direction, streamlining and focusing the direction of research on climte change in late 1988. Prior to that time, research on global warming was taking place in several different federal agencies. As a result, research tended to be scattered and duplicative. Officials decided a more efficient approach would be to coordinate all research programs through a single agency, now called the U.S. Global Change Research Program. This program's four major objectives are to determine how the Earth's environment has changed in the past, what forces are responsible for global change, how the Earth responds to these forces, and how global change can be predicted. Similar efforts are being planned on an international level. The International Council of Scientific Unions is organizing an International Geosphere-Biosphere Program for the early 1990s. The purpose of this program is to coordinate efforts of scientists from around the world in research on global change.

A third position is that action can and must be taken, as soon as possible, to reduce emissions of carbon dioxide and other greenhouse gases. Those who take this position recognize the financial and social costs involved with this proposal. But they argue that the stakes are much too high to wait or to do nothing. They argue that, even with immediate action, warming trends are likely to continue well into the next century.

A number of factors have contributed to the growing concern about global warming. First, many scientists are convinced that reliable climate trends have begun to appear. Dependable data on carbon dioxide levels in the atmosphere, global temperatures, and other important variables have been available only since the late 1950s. In the 1960s and 1970s, it was still too soon to tell whether data represented long-term *climatic trends* or short-term *weather patterns*. By the 1980s and 1990s, scientists had a better basis for choosing between these options, and more and more of them decided that long-term trends were becoming apparent.

The case for immediate action was strengthened by weather patterns of the 1980s. Ordinary people had been hearing about the greenhouse effect for a decade. Then, along came the warmest decade in recorded history. Everything scientists had been warning about appeared to be taking place. Scientists knew weather patterns for a single decade meant very little in terms of climatic change. But, because of those patterns, popular opinion suddenly began to favor a more serious, more aggressive consideration of global warming.

Second, scientists began to realize that gases other than carbon dioxide were involved in global warming and that the concentration of these gases was increasing also, sometimes much more quickly than that of carbon dioxide. The problem seemed troubling because we tend to know even less about the origin and effects of methane, nitrous oxide, CFCs, and other greenhouse gases than we do about carbon dioxide.

Third, studies on ozone depletion convinced many authorities of the damage that humans can do to the atmosphere. The *ozone layer* is a thin region of the stratosphere that is especially rich in ozone. Studies have shown that the concentration of ozone in this layer is gradually decreasing. Many scientists believe synthetic products released from the Earth's surface, especially CFCs, are responsible for this change. In 1985, British scientists reported a sudden large decrease in the concentration of ozone in the stratosphere over the Antarctic. They referred to the condition as a "hole" in the ozone layer. Each year since 1985, the "hole" has become larger (Patrusky 1988).

The appearance of the ozone hole has had a profound impact on scientists. Theories had been able to explain a modest and gradual loss of ozone, but the nearly total loss of ozone in large regions of the stratosphere came as a shock. The phenomenon suggests that human activities may produce more extensive effects on the atmosphere in shorter time periods than anyone had imagined possible. If that is the case, global warming may already be occurring more quickly than anyone had considered possible.

Those who call for immediate action on global warming understand the practical obstacles they face. Dealing with a global issue such as climate change could well cost billions of dollars and demand significant changes in lifestyle. Therefore, members of this camp sometimes try to make their appeal for action more palatable. They look for changes that would be environmentally desirable even if there were no global warming. For example, reduced dependence on fossil fuel energy in the United States and other developed nations would have both political and health benefits. By setting standards for reduced greenhouse gas emissions, we could achieve a number of desirable goals, a reduction in global warming being only one.

Those who argue for immediate action focus on three options: technological fixes, political action, or both. The term *technological fix* refers to some scientific technique or method for counteracting an undesirable effect, the effect in this case being an increase in global temperatures. One example of a technological fix for global warming is the proposal to add gases to the upper atmosphere. These gases would increase the reflectivity of the atmosphere and reduce the amount of solar energy reaching the Earth. Perhaps the cooling effect thus produced could compensate for some or all of any global warming that occurs.

Another technological fix is to add iron powder to the world's oceans. Studies suggest that phytoplankton—one-celled ocean-dwelling organisms—grow much more rapidly when iron is included in their diet. Because phytoplankton are a major consumer of carbon dioxide, increasing their population might result in a dramatic decrease in carbon dioxide in the atmosphere and the Earth's oceans.

One problem with technological fixes is that they tend to require massive delivery programs. For example, dumping gases into the upper atmosphere would require the use of hundreds of space shuttle–type vehicles, a project whose expense and technical challenges would be staggering. Technological fixes also hold the potential for creating huge new problems. No one really knows what might happen, for example, if tons of iron powder were sprinkled on top of the Earth's oceans. As a result, radical experiments such as the ones described above are often regarded by scientists and nonscientists alike as hopelessly absurd or unrealistic. They probably would never be attempted until and unless both scientists and government officials were convinced that extreme methods had to be used to deal with the accumulation of greenhouse gases in the atmosphere.

Far more common are proposals for some sort of political action that will reduce the emission of carbon dioxide and other greenhouse gases. One of the most significant actions of this kind was the 1992 United Nations Conference on Environment and Development held in Rio de Janeiro, Brazil. One objective of that conference was to develop a strategy by which the nations of the world could reduce the emission of greenhouse gases worldwide.

Obtaining consensus among the 160 participant nations, however, was no easy matter. In the first place, controlling emissions is a far different problem for developed nations than it is for developing nations. The average U.S. citizen, for example, is responsible for the emission of 5 tons of carbon dioxide each year, 12 to 50 times the average amount produced by a citizen of India or China. Developing nations tend to ask why *they* should be asked to cut back on the industrial production that produces carbon dioxide emissions when such production is their best hope for progress. It is hypocritical, they say, for developed nations to suggest that developing nations should not try to follow the same path that they (the developed nations) once took themselves.

In fact, most developed nations at Rio appeared willing to accept fairly severe standards to limit carbon dioxide emissions. An early draft document called for cutting back emissions to 1990 levels by the year 2000.

The only nation to oppose that plan was the United States. President Bush's argument was that too little was known about global warming to justify such radical actions. It would not be fair to ask business and industry to cut back on their operations, he said, when so many questions still remain about climate change. Money could be better spent on further research and on economic development, he suggested. In a compromise agreement, the final Rio statement included a more general commitment, calling only for "the return by the end of the decade to earlier levels" of greenhouse gas emissions (Miller 1992).

What Can Be Done

If global warming is a real possibility, the most obvious immediate solution is to reduce carbon dioxide emissions. In turn, that means reducing the world's dependence on fossil fuels. For example, the 1988 International Conference on the Changing Atmosphere, held in Toronto, Canada, adopted a resolution calling for a 20 percent cut in carbon dioxide emissions by the year 2005 ("The Changing Atmosphere" 1989).

Recommendations for achieving this kind of reduction usually take one of two forms: increased conservation or development of alternative energy sources. Critics point out that, by any standard, the United States is the single largest producer of carbon dioxide in the world. Each year, Americans release about 1 billion tons of carbon dioxide into the atmosphere, one-fifth of the world total. No plan for dealing with global warming is likely to succeed until and unless the United States cuts back on its consumption of coal, oil, and natural gas.

One step in that direction would be for U.S. citizens and industry to start using fossil fuels more carefully. On average, each Japanese citizen uses only half as much energy as does the average American. Yet the Japanese standard of living is hardly inferior to that of Americans. The main difference is that the United States has always had an abundant supply of fossil fuels (and, thus, little motivation to conserve) whereas Japan has virtually no fuel resources of its own.

Conservation advocates point out any number of ways in which the United States (and other developed nations) can use energy more efficiently. They explain that each time a homeowner replaces a 75-watt incandescent bulb with an 18-watt fluorescent bulb, 180 kilograms (400 pounds) of coal are saved annually. They say that improved insulation in homes and office buildings could reduce energy use in the United States by 50 to 75 percent. They argue that improving the efficiency of automobile engines and increasing the use of public transportation could reduce carbon dioxide emissions by half.

Another step toward reducing our dependence on fossil fuels would be to develop alternative sources of energy, such as wind power, geothermal energy, solar power, and nuclear energy. Scientists at the U.S. Solar Energy Research Institute, for example, estimate that half of all U.S. electrical needs could be met by photovoltaics (solar power) by the middle of the twenty-first century (U.S. Congress, Office of Technology Assessment 1991).

Nuclear power is perhaps the most immediately available alternative to fossil fuels. Worldwide, increased use of nuclear power since 1975 has resulted in an annual decrease of nearly 300 million tons of carbon emissions, about 5 percent of the annual total. But safety concerns have all but eliminated nuclear power as an alternative in the United States.

An interesting new view on efforts to control global warming was revealed in a 1990 World Resources Institute (WRI) report. Traditionally, responsibility for controlling greenhouse effects has been laid at the door of developed nations. Critics have pointed to motorized transportation and industrial processes as the major sources of carbon dioxide emissions and, therefore, the primary targets of emission control programs.

In their 1990 report, however, WRI included three developing nations—Brazil, China, and India—among the world's five largest contributors to global warming problems. (The United States and the former Soviet Union were the other two worst "villains.") Brazil earned its place on the list primarily because of its massive programs of forest burning. These burnings not only release enormous amounts of carbon dioxide into the air but also remove an important sink for carbon dioxide: the trees themselves.

China and India received high rankings not because of carbon dioxide emissions, but because of methane released by agricultural processes. Cultivation of rice and cattle raising in these two countries result in levels of methane emissions that are beginning to rival the production of carbon dioxide in developed nations. How these contributions to global warming can be brought under control is not at all clear (Hammond, Rodenburg, and Moomaw 1991).

Conclusion

If any one word could be used to describe the problem of global warming, it might be *uncertainty*. Data about greenhouse gas emissions, temperature changes, and possible climate effects are all characterized by uncertainty. So are recommendations for ways of dealing with the emission of greenhouse gases. Given all these uncertainties, it seems likely that the debate over global warming will continue well into the future.

References

Begley, Sharon. 1989. "Is It All Just Hot Air?" *Newsweek* (November 20).

Boden, Thomas A., Robert J. Sepanski, and Frederick W. Stoss. 1991. *Trends '91: A Compendium of Data on Global Change*. Oak Ridge, TN: Carbon Dioxide Information Analysis Center.

"The Changing Atmosphere: Implications for Global Security." 1989. In *The Challenge of Global Warming*, edited by Dean Edwin Abrahamson. Washington, DC: Island Press.

Edgerton, Lynne T. 1991. *The Rising Tide: Global Warming and World Sea Levels*. Washington, DC: Island Press.

Hammond, A. L., Eric Rodenburg, and William R. Moomaw. 1991. "Calculating National Accountability for Climate Change." *Environment* (January/February).

Hileman, Bette. 1989. "Global Warming," *Chemical & Engineering News* (March 13).

Houghton, R. A., et al. 1989. "The Flux of Carbon from Terrestrial Ecosystems to the Atmosphere in 1980 Due to Changes in Land Uses: Geographic Distribution of Global Flux." In *Slowing Global Warming: A Worldwide Strategy*, p. 29. Edited by Christopher Flavin. Washington, DC: Worldwatch Institute.

Intergovernmental Panel on Climate Change. 1992. "1992 IPCC Supplement." [photocopied report] (February).

Lindzen, R. S. 1990. "A Skeptic Speaks Out." *EPA Journal*. (March/April).

Lippsett, Laurence. N.d. "Wallace Broecker '53: The Grandmaster of Global Thinking." *Columbia College Today*.

Miller, Stanton S. 1992. "The Road from Rio." *ES&T*. (September).

Patrusky, Ben. 1988. "Dirtying the Infrared Window." *Mosaic*. (Fall/Winter).

Salati, E., R. L. Victoria, L. A. Martinelli, and J. E. Richey. 1991. "Forests: Their Role in Global Change, with Special Reference to the Brazilian Amazon." In *Climate Change: Science, Impacts, and Policy*. Edited by J. Jäger and H. L. Ferguson. Cambridge: Cambridge University Press.

Seidel, Stephen, and Dale Keyes. 1983. *Can We Delay a Greenhouse Warming?* 2d ed. Washington, DC: Office of Policy Analysis, Office of Policy, Planning and Evaluation.

Smith, J. B., and D. Tirpak, eds. 1990. *The Potential Effects of Climate Change on the United States.* 3 vols. New York: Hemisphere Publishing.

Thompson, L. G., et al. 1989. "Holocene-Late Pleistocene Climatic Ice Core Records from Qinghai-Tibetan Plateau." *Science.* (October 27).

U.S. Congress, Office of Technology Assessment. 1991. *Changing by Degrees: Steps To Reduce Greenhouse Gases.* OTA-O-482. Washington, DC: U.S. Government Printing Office.

U. S. Government Printing Office. 1988. "Greenhouse Effect and Global Climate Change." Hearing before the U.S. Senate Committee on Energy and Natural Resources, June 23, 1988. Washington, DC: U. S. Government Printing Office.

Woods, J. D. 1984. "The Upper Ocean and Air-Sea Interactions." In *The Global Climate.* Edited by J. T. Houghton. Cambridge, England: Cambridge University Press.

2

Chronology

2.5 million years before the present (B.P.)
First appearance of the polar ice caps. Prior to this time, the Earth's average annual temperature was apparently high enough (16° to 23° C; 60° to 73° F) to prevent the formation of large, long-lasting masses of ice anywhere on the planet. Since this date, the Earth has experienced alternative periods of cooling and warming. During the periods of cooling, which have lasted anywhere from 50,000 to 80,000 years, glaciers have covered large parts of North America, Europe, and Asia. These glaciers have receded during the shorter interglacial periods (lasting about 10,000 years), when temperatures have risen. The Earth has experienced seven ice ages since about 700,000 B.P.

18,000 B.P.
The most recent ice age reaches its peak. The Earth's average annual temperature falls to about 10° C (50° F), probably the coldest in the planet's history.

11,000 B.P.
The current interglacial period begins. Average temperatures during this period are about 15° C (59° F). This period is characterized by periodic warming and cooling trends during which temperatures change by no more than a few degrees one way or the other. During one warm period, known as the Big Climatic Optimum (about 6000 B.P.), summer temperatures increase by an average of about 2° C (3.6° F). A second warming trend, the Little Climatic Optimum, occurs between about A.D. 800 and 1250. During this period, the Earth's average

11,000 B.P.
cont.

annual temperature is about 1° C (1.8° F) higher than the Earth's present average summer temperature. Polar ice sheets retreat to an extent that many northern regions develop temperate climates. Settlers arrive in Greenland and name the island for the lush vegetation that exists there.

About A.D. 1400 to 1700

The Little Ice Age arrives. Average temperatures drop by about 1° C (1.8° F) to about 14° C (57° F). Polar ice sheets move southward once more. Greenland is largely covered by ice, and most human communities die out. As the Little Ice Age comes to a close, annual average temperatures return to their previous level of about 15° C (59° F) and remain essentially constant until the present day.

1753

Joseph Black discovers carbon dioxide by treating limestone (calcium carbonate) and "magnesia alba" (magnesium carbonate) with acids. Black gives the name "fixed air" to the gas he discovers. He later finds that "fixed air" is present in the atmosphere, is produced during the fermentation of beer, and is contained in air exhaled by humans.

1769

James Watt "invents" the steam engine. He actually improves on an existing device originally developed by Thomas Newcomen in about 1712. But Watt's improvement is such an advance that Newcomen's earlier work is eventually forgotten or ignored. The steam engine is widely regarded as the single most important factor in the rise of the Industrial Revolution.

1827

Jean-Baptiste Fourier, a famous French mathematician, outlines a process by which solar energy is captured by the Earth's atmosphere, thus raising the planet's temperature. He suggests the name *greenhouse effect* for this phenomenon as an analogy (one that is not entirely correct) to the way glass windows allow for the warming of the inside of a greenhouse.

1861–1863

The English physicist John Tyndall studies the absorption of infrared radiation by carbon dioxide and water. He concludes that a decrease in the concentration of carbon dioxide in the atmosphere may result in the development of an ice age.

1902   Léon Philippe Teisserene de Bort announces that the Earth's atmosphere consists of two layers. Six years later he names the lower layer the *troposphere*. Teisserene de Bort's discoveries are made by using unmanned balloons that contain clock-driven equipment that continuously records temperature, pressure, and other atmospheric variables. Teisserene de Bort's balloons, made of varnished paper, rise to a height of almost 16 kilometers (10 miles).

1908   The Swedish chemist Svante Arrhenius argues that carbon dioxide molecules in the Earth's atmosphere act as a means of trapping heat reflected from the Earth's surface. His theory is the first clear explanation of the greenhouse effect.

1920   Milutin Milankovitch, a Serbian engineer, publishes a first book explaining how variations in the Earth's position in space explain variations in the amount of solar radiation reaching the planet's surface.

1938   The British engineer G. D. Callender attempts to show how carbon dioxide released by human activities (anthropogenic carbon dioxide) has affected atmospheric temperature. He is the first scientist to describe in detail how anthropogenic carbon dioxide has influenced climate in the past and to what extent it may do so in the future. He regards these changes as beneficial for the future of agriculture in the North Temperate Zone and as a safeguard against the return of another ice age.

1940–1965   A modest cooling period occurs over much of the Northern Hemisphere. Although the average temperature decline is no more than about 0.3° C (0.5° F) overall, some regions experience winters that are up to 2° to 3° C (4° to 5° F) cooler. Some writers talk about the beginning of a new ice age on the Earth.

1950   Jule G. Charney leads a team of scientists in developing the first computer program designed to model weather and climate changes. The program runs on a primitive ENIAC computer operated by the U.S. Army Signal Corps. Charney works under the direction of John von Neumann at the Institute for Advanced Study and is later called "the father of numerical weather prediction."

1950
cont.

The first effort to model the global atmosphere is developed by Norman Phillips in 1956.

1957

Charles Keeling, a postdoctoral student at the California Institute of Technology, initiates the longest continuous series of detailed atmospheric measurements in modern history. He establishes monitoring stations on Mauna Loa in Hawaii and at the South Pole to sample the concentration of carbon dioxide in the atmosphere.

Roger Revelle and Hans Suess of Scripps Institute of Oceanography warn in an article in the journal *Tellus* (volume ix, number 1) that excess emissions of carbon dioxide from human sources are not being absorbed by the oceans. They warn that "mankind in spite of itself is conducting a great one-time geophysical experiment."

**July 1957
to
December
1958**

An 18-month period designated as the International Geophysical Year (IGY) during which scientists from 66 nations carry out a worldwide research effort under the coordination of the International Council of Scientific Unions (ICSU). The IGY is a program designed to make detailed observations about such phenomena as terrestrial magnetism, cosmic radiation, and atmospheric conditions.

1963

The National Oceanic and Atmospheric Administration (NOAA) establishes the Geophysical Fluid Dynamics Laboratory (GFDL) at Princeton University. The laboratory, under the direction of Joseph Smagorinsky, is devoted to a mathematical modeling of the atmosphere.

1967

Wallace Broecker of the Lamont-Doherty Geological Observatory uses a computer simulation to predict the effects of doubling the amount of carbon dioxide in the atmosphere. He finds that such a change will result in an increase of 2.4° C (4.3° F) in the planet's average annual temperature.

1974

Mario Molina and F. Sherwood Rowland, two chemists at the University of California at Irvine, predict that the family of synthetic chemicals known as chlorofluorocarbons (CFCs) may result in the loss of up to 20 percent of the atmosphere's ozone layer. Although some scientists are convinced by the Molina-Rowland argument, others doubt that human activities can have such far-reaching effects on the Earth's atmosphere.

1976   James Hays of the Lamont-Doherty Geological Observatory, John Imbire of Brown University, and Nicholas Shackleton of the University of Cambridge devise a method of studying changes in ice cover over a half million years. They study variations in isotope ratios found in ice cores.

1977   David Slade of the Energy Research and Development Administration (later, the Department of Energy), calls a meeting of scientists to discuss a possible research program to study the role of carbon dioxide in the environment.

1979   The United Nations (UN) World Meteorological Organization (WMO) sponsors the first World Climate Conference in Geneva, Switzerland. The conference considers issues such as the greenhouse effect, climate change, drought, and soil erosion.

Frank Press, science advisor to President Jimmy Carter, asks the National Academy of Sciences (NAS) to conduct a study of the greenhouse effect. The NAS panel, chaired by Jule Charney, concludes that a doubling of carbon dioxide in the atmosphere from anthropogenic sources would raise global temperatures by $3°$ C $\pm 1.5°$ C ($5.4°$ F $\pm 2.7°$ F).

Oceanographer Roger Revelle, ecologist George Woodwell, geophysicist Gordon MacDonald, and Charles Keeling, founder of the Mauna Loa carbon dioxide monitoring project, report to the Council on Environmental Quality (CEQ) that "man is setting in motion a series of events that seem certain to cause a significant warming of world climates unless mitigating steps are taken immediately."

The U.S. Department of Energy (DOE) sponsors a meeting in Annapolis, Maryland, to discuss the question of what the effects on human society might be if climate changes do occur. In addition to scientists, DOE invites economists, historians, anthropologists, political scientists, and psychologists to the meeting.

1980   Ronald Reagan is elected president of the United States. Over the next dozen years, the administrations of Reagan and his successor, George Bush, take a much more cautious approach to environmental problems in general and to global warming in particular than had their immediate predecessors.

1981   The CEQ releases a report concluding that "the potential risks from even moderate increases in the burning of fossil fuels . . . underscore the vital need to incorporate the CO_2 issue into the development of United States and global energy policy."

Scientists at the National Aeronautics and Space Administration's (NASA) Goddard Institute of Space Studies (GISS) point to the role of methane, nitrous oxide, ozone, and CFCs as greenhouse gases.

A report by James E. Hansen and other scientists at GISS, published in the August 28 issue of the journal *Science,* summarizes trends in mean global temperatures from 1880 to 1980. The report concludes that there is a "high probability of [global] warming in the 1980s." Some authorities call this report the first clear evidence for the existence of global warming patterns.

1983   A group of scientists meet in July at Woods Hole, Massachusetts, to consider an IGY-style geosphere-biosphere research program. One result of the meeting is the creation of a committee for what is to become the International Geosphere-Biosphere Program (IGBP). A formal proposal for such a program is also submitted to the ICSU.

NASA creates an Earth Systems Sciences Advisory Committee to suggest a program for studying global change. The committee will submit a report in 1986 recommending a unified study of global change involving joint participation of NASA, NOAA, the National Science Foundation (NSF), and other governmental agencies. The committee report will suggest that the study is needed because "a better understanding of the earth and its immediate environment is essential if we are to improve our ability to detect and respond to warnings of significant global change."

The Environmental Protection Agency (EPA) issues its first report on global warming, "Can We Delay a Greenhouse Warming?" The report concludes that carbon dioxide levels in the atmosphere are already sufficient to ensure a rise in global temperatures of about 2° C (3.6° F) and that economic trends will ensure a continuing increase in carbon dioxide concentrations and global temperatures for the near future.

The National Academy of Sciences (NAS) issues a report on global warming with quite different conclu-

1983
cont.

sions from those of the EPA. NAS is *not* of the opinion that "the evidence at hand about CO_2-induced climate change would support steps to change current fuel use patterns away from fossil fuels."

1985

Veerabhadran Ramanathan of the University of Chicago and Ralph Cicerone of the National Center for Atmospheric Research (NCAR) warn that greenhouse gases other than carbon dioxide are at least as important in trapping reflected heat in the atmosphere as is carbon dioxide. Ramanathan's calculations show that present emission trends will result in the appearance of a greenhouse effect about 50 years sooner than as the result of carbon dioxide alone.

An international meeting of scientists from 29 countries, sponsored by the UN Environmental Program (UNEP) and WMO in Villach, Austria, concludes that "some warming of climate now appears inevitable." It points out, however, that "the rate of future warming could be profoundly affected by government policies on energy conservation, on use of fossil fuels, and emission of some greenhouse gases."

U.S. Senator Albert Gore calls for an international "Year of the Greenhouse."

British scientists report a 40 percent decrease in the concentration of ozone over the Antarctic. The phenomenon will continue to be observed in succeeding years with the ozone loss increasing annually. Later evidence will suggest that the "ozone hole" goes back at least as far as 1974. The discovery of the ozone hole is significant because it illustrates the dramatic, large-scale, and somewhat unexpected effect that human activities can have on the Earth's atmosphere.

1986

The NSF recommends the creation of a Global Geosciences Initiative that includes seven areas of study, varying from solid-earth dynamics to tropospheric chemistry.

U.S. Senator John Chafee holds hearings on the greenhouse effect. At these hearings, Dr. James Hansen of GISS suggests that significant warming trends may be detectable within a decade or two. Chafee asks the EPA and the Office of Technology Assessment to develop policy options for stabilizing the level of greenhouse gases in the atmosphere.

1986
cont.

A three-volume report on atmospheric ozone, produced by scientists at NASA, WMO, and other agencies, documents the speed with which atmospheric changes are taking place as a result of human activities.

1987

A team of Soviet and French scientists take an ice core 2,000 meters (6,600 feet) deep at Vostok in the Antarctic. By analyzing air bubbles trapped in the ice, they are able to estimate atmospheric composition and temperature over a period of about 160,000 years.

The U.S. Office of Science and Technology Policy establishes the interagency Committee on Earth Sciences, whose purpose is to "coordinate the development and implementation of an integrated Federal research program which would allow the United States to address the issues posed by current and anticipated changes in the global environment." Funding for the resulting U.S. Global Change Research Program is first included in the president's budget for fiscal year 1990.

The Senate Environment and Public Works Committee holds hearings on global warming. Veerabhadran Ramanathan testifies that human activities up to 1980 had already committed the Earth to a 0.7° to 2.0° C (1.3° to 3.6° F) temperature increase.

On September 16, 24 nations sign the Montreal Protocol on Substances That Deplete the Ozone Layer. The protocol calls for a 50 percent reduction in CFC consumption by 1999.

Gordon MacDonald, vice-president and chief scientist at the MITRE Corporation, calls for a carbon tax before the Senate Energy Committee. The purpose of a carbon tax is to discourage the use of carbon-based fuels, the combustion of which releases carbon dioxide to the atmosphere. At the same hearing, Gus Speth, president of the World Resources Institute, suggests that Congress pass comprehensive legislation to combat global warming. Speth's suggestion forms the basis of legislation that will be introduced by Senator Tim Wirth in 1988.

At their summit meeting in December, President Reagan and General Secretary Mikhail Gorbachev of the Soviet Union agree to continue joint U.S.-Soviet studies on climate change.

The Governing Council of UNEP meets in Nairobi, Kenya, and agrees to begin the search for ways of deal-

1987
cont.

ing with the problem of global warming. The organization also decides to create, with the WMO, an intergovernmental body to conduct ongoing studies of global warming. That organization, the Intergovernmental Panel on Climate Change (IPCC), is officially established in November 1988. IPCC is now the most important single agency dealing with climate change on an international level.

1988

In January, President Reagan signs the Global Climate Protection Act, which requires, among other provisions, that the president submit to Congress a plan for stabilizing the concentration of greenhouse gases in the atmosphere.

The EPA publishes a three-volume report entitled *The Potential Effects of Climate Change on the United States.* The report acknowledges that existing information on such effects is very incomplete, but that some effort should be made to develop various scenarios that might result from global warming.

A joint UNEP/WMO report issued in April warns that the pace of climate change is increasing at a rate to which natural systems cannot adapt. The report suggests that actions be taken to reduce the emission of greenhouse gases. It also outlines issues to be considered at the forthcoming Toronto conference on global change.

An international conference, "The Changing Atmosphere: Implications for Global Security," is held in Toronto, Canada, from June 27 to 30. The conference calls for a reduction in carbon dioxide emissions by about 20 percent of current levels by the year 2005. The special problems faced by developing nations in meeting this goal are recognized and the establishment of a World Atmosphere Fund is proposed as a way of dealing with these problems. The fund is to obtain its money from a carbon tax on developed nations.

During his fall campaign for the presidency, George Bush promises, if elected, to conduct a vigorous campaign to deal with global warming.

British Prime Minister Margaret Thatcher first addresses the issue of global warming in a speech before the Royal Society on September 27.

The United Kingdom's Meteorological Office announces that its global circulation model predicts that

1988
cont.

a doubling in carbon dioxide concentration will result in an increase of 5.2° C (9.4° F) in global temperatures, a change that could occur as early as 2050.

Testifying before the Senate Energy Committee on June 23, James Hansen announces that he is "99 percent confident" that the greenhouse decade was not a random event, but a real indication of global climate change. "Global warming," he says, "is now sufficiently large that we can ascribe with a high degree of confidence a cause and effect relationship to the greenhouse effect" and that "extreme events such as summer heat waves and . . . heat wave/drought occurrences in the Southeast and Midwest United States may be more frequent in the next decade." Hansen's remarks spark an immediate and controversial debate among scientists and politicians about global warming. Many of his colleagues feel Hansen has overstated the case for global warming, but all agree that he has drawn public attention to the issue as never before.

An article in the December 23 issue of *Science* cites work by Tom Karl of the U.S. National Climate Data Center to the effect that "there may have been no global warming to speak of during the last century." Karl's report is widely misinterpreted to mean that he personally does not believe that global warming is a reality.

1989

A report published by scientists at the George C. Marshall Institute raises doubts about global warming. It concludes that "it is possible that a combination of natural and solar variability is the cause of the entire temperature increase observed since 1880, with the greenhouse effect relegated to a negligible role." A number of scientists disagree with the Marshall report. The director of NOAA's Geophysical Fluid Dynamics Laboratory, for example, writes in the journal *Science* that "there are uncertainties, but I can't think of any combination of them that could conspire to make the [greenhouse] problem go away." The Marshall report appears, nonetheless, to be very influential in the Bush administration.

Two weeks before the inauguration of George Bush as president, the U.S. National Academy of Sciences (NAS) recommends to him that global warming be placed high on his agenda. NAS warns that "the future welfare of human society" is at risk.

1989
cont.

A study by Kirby Hanson of NOAA finds that no change in U.S. temperature or rainfall can be documented in the period 1895 to 1988. Hanson points out that these results do not mean that the mechanisms required to bring about global warming have not already been put into motion.

A working group of the IPCC meets in Washington, D.C., to begin planning for ways to deal with global warming. The group rejects a draft U.S. proposal on the grounds that it will delay actions needed to deal with the problem.

A March 11 "environmental summit" in The Hague, The Netherlands, brings together presidents, prime ministers, and other government officials from 24 nations. The group recommends the creation of a new agency within the United Nations to "be responsible for combating any further global warming of the atmosphere." To avoid political conflicts between the world's two superpowers, the United States and the Soviet Union are not invited to the conference.

Meeting in May at Helsinki, Finland, 80 nations agree to amend the 1987 Montreal Protocol to phase out *all* use (rather than 50 percent) of CFCs by 2000.

Members of the Bush administration order James Hansen of NASA to change his congressional testimony on global warming to suggest that scientists still do not know about the effects of anthropogenic greenhouse gases on climate change.

The Bush administration first announces that it will not work on a proposed UN greenhouse treaty because too little is known about the economic impact of global warming, and then does a turnabout and agrees to participate in such an effort.

The President's Office of Science and Technology announces a 10-year research program to find out "how this planet ticks." The president asks for $191.5 million for the program for the 1990 fiscal year.

An international conference of 68 nations on global warming unanimously adopts a resolution committing these nations to stabilizing carbon dioxide emissions by 2000. The resolution does not set the levels at which stabilization will occur.

1980s

A period sometimes known as "the greenhouse decade" because the Earth's annual average temperature is the

1980s
cont.

warmest ever recorded by scientists for any 10-year period. The six warmest years in modern history—1980, 1981, 1983, 1987, 1988, and 1989—occur during this decade.

1990

An analysis by the President's Council of Economic Advisors says that switching to less polluting energy systems as a way of dealing with global warming could cost "$800 billion under optimistic scenarios of available fuel substitutes and increasing energy efficiency to $3.6 trillion under pessimistic scenarios between now and 2010."

The Union of Concerned Scientists presents a petition to President Bush urging him to take action on global warming. The petition is signed by 49 Nobel laureates, 64 recipients of the National Medal of Science, and more than 700 members of the NAS.

Speaking to a meeting of the IPCC, President Bush expresses concern about global warming but claims that more research is needed on the problem before action can be taken.

A study by British scientists commissioned by the United Nations finds that global warming is "a virtual certainty."

The United States hosts a conference on global warming on April 17–18 in Washington, D.C. President Bush calls for further study on the problem, but representatives from the European Economic Community (EEC) argue that "gaps in knowledge must not be used as an excuse for inaction" and that "Americans are falling behind on this, and . . . the time for action has come."

A May 2 conference of legislators from 42 countries calls for a "global Marshall plan for sustainable development and the environment."

The United States refuses to contribute $25 million over a three-year period to help developing nations reduce their use of CFCs.

More than 130 nations meet in Geneva, Switzerland, for the Second World Climate Conference. The conference declaration calls for all nations to begin setting targets or establishing programs for reducing the emission of greenhouse gases. The setting of specific targets is blocked by opposition from the United States and the Soviet Union. The conference calls for a February 1991 meeting in Washington, D.C., to find ways of imple-

1990
cont.

menting this objective and a June 1992 international convention on global warming. The international convention will eventually become known as the "Earth Summit." At the conference in Geneva, Japan announces plans to stabilize carbon dioxide emissions by 2000. Germany announces a goal of reducing carbon dioxide emissions by more than 25 percent by 2005. Similar plans had been announced earlier by other nations of the EEC.

1991

Meteorologists in the United States and the United Kingdom report that 1990 was the warmest year in recorded history.

In an April 10 report, the National Academy of Sciences suggests a number of energy conservation measures that would reduce greenhouse gas emission by up to 40 percent over the next 30 years.

In a first for U.S. utility companies, the Los Angeles Department of Water and Power and Southern California Edison pledge to reduce carbon dioxide emissions by 20 percent by 2010.

A report in the journal *Nature* on July 4 announces that the Arctic ice cap decreased in size by 2 percent between 1978 and 1987.

Worldwatch Institute reports on December 8 that worldwide emissions of carbon dioxide dropped by about 0.2 percent, from 5.813 billion tons to 5.803 billion tons, during 1990.

1992

UN-sponsored talks designed to produce a global warming treaty meet with limited success. A critical issue involves the question of providing aid to developing nations to allow them to reduce greenhouse gas emissions without hindering their own development.

On March 24, President Bush announces that he will not sign any agreement on global warming that would "throw a lot of Americans out of work." He refuses to say whether he will attend the Earth Summit planned for Rio de Janeiro, Brazil, in June. UN official Cado Ripa di Meana calls Bush's position "an attack at the very heart of the conference."

Representatives from 143 nations approve a draft treaty on greenhouse gas emissions in a May 9 meeting in New York City. The treaty is to be presented to world leaders at the Earth Summit in Rio in June. As a

1992
cont.

concession to U.S. concerns, the final version of the treaty makes no mention of specific emission targets. Three days later, President Bush announces that he will attend the Rio meeting.

The 1992 energy bill passed by Congress includes a number of provisions designed to reduce carbon dioxide emissions. One such provision sets aside $15 billion over a five-year period for research on alternative energy sources.

The UN Conference on Environment and Development (the "Earth Summit") is convened in Rio de Janeiro from June 3 to 14. Representatives from 178 nations meet to discuss ways of protecting the global environment without stifling economic growth in developing or developed nations. Specific agreements deal with biodiversity, environmental cleanup strategies, forest preservation, sustainable development, and global warming. The United States is criticized by many conference participants for its refusal to sign the biodiversity treaty and its efforts to "water down" the global warming agreement.

3

Biographical Sketches

HOW THE EARTH'S CLIMATE CHANGES OVER TIME has intrigued scientists for at least 100 years. Some of the greatest names in chemistry, physics, geology, astronomy, and other fields of science contributed to early theories of the greenhouse effect and global warming. Only recently, however, has climate change become a specialized field of research to which a person could devote his or her whole career. This chapter includes brief biographical sketches of some of the most important scientists, past and present, who have contributed to this field. In addition, the chapter contains biographical information about figures from business, politics, and other fields who have had an impact on climate change policy.

Svante August Arrhenius, 1859–1927

Arrhenius was born in Wijk, Sweden, on February 19, 1859. He was a child prodigy who taught himself to read at the age of three. His doctoral thesis at the University of Uppsala dealt with the theory of ionization, a theory that earned him the Nobel Prize for Chemistry in 1903.

In 1891 Arrhenius was appointed to a teaching position at the University of Stockholm. He remained at Stockholm until 1905, when he was appointed director of the Nobel Institute for Physical Chemistry. He died in Stockholm on October 2, 1927, shortly after retiring from the Nobel Institute.

Arrhenius was interested in a great variety of scientific topics. In 1908, for example, he published *Worlds in the Making,* a book that outlined a theory showing how human life may have originated from spores that arrived on Earth from outer space. In this book, he analyzed the process by which carbon dioxide in the Earth's atmosphere traps heat and makes our planet hospitable for human habitation. He calculated the temperature changes that would occur if the concentration of carbon dioxide were to increase or decrease by certain amounts. Arrhenius's calculations of a 5° to 6° C (9° to 11° F) increase from a doubling of carbon dioxide concentration are very close to those made by modern scientists.

Wallace S. Broecker, 1931–

"The grandmaster of global thinking" is a title bestowed on Wallace Broecker by one of his colleagues at Columbia University. Another colleague has referred to the university's prestigious Lamont-Doherty Geological Observatory as "Wally Broecker, Inc."

For the better part of four decades, Broecker has been trying to untangle the complex interaction of atmospheric gases, the circulation of ocean water, and the Earth's climate. He has come to the conclusion that climatic change in the past has sometimes come about very suddenly and without warning.

Broecker was born in Chicago, Illinois, on November 29, 1931. He received his bachelor of arts (1953) and Ph.D. (1958) degrees from Columbia University. After graduation, he was appointed assistant professor at Columbia, where he has remained ever since. Since 1977 he has been Newberry Professor of Geology at Columbia. Broecker is the author or coauthor of three scholarly textbooks and of one book for undergraduates that he wrote and published at his own expense in 1987, *How To Build a Habitable Planet.*

George Herbert Walker Bush, 1924–

As president of the United States from 1988 to 1992, George Bush was largely responsible for shaping the nation's official policy about global warming. Essentially, Bush's philosophy was to move slowly on the problem of greenhouse gases. He believed that taking action to deal with greenhouse gases could cost the nation billions of dollars, which he viewed as an enormous investment on an issue about which so much controversy remained.

George Bush was born in Milton, Massachusetts, on June 12, 1924. He attended the Greenwich Country Day School, Phillips Academy, and Yale University, from which he received his bachelor's degree in economics in 1948. His education was interrupted in 1943 when he joined the U.S. Navy Reserve. After receiving his commission, he became the youngest pilot in the navy.

Following graduation from Yale, Bush declined an opportunity to join his father's firm on Wall Street and moved to Texas instead. There he cofounded the Bush-Overby Company, which dealt in oil and gas properties.

Bush served as U.S. representative from Texas's 7th congressional district (1966 to 1970), U.S. permanent representative to the United Nations (1970 to 1973), chairman of the Republican Party (1973 to 1976), chief liaison officer to China (1976 to 1977), and director of the Central Intelligence Agency (1977 to 1980). In 1980 and 1984 he was elected vice-president under Ronald Reagan. He was defeated in his presidential reelection bid by Bill Clinton in 1992.

Thomas Chrowder Chamberlin, 1843–1928

Chamberlin was born in Mattoon, Illinois, on September 25, 1843. He received his bachelor's degree from Beloit College and then did graduate work at the universities of Michigan and Wisconsin. He became professor of geology at Beloit in 1873 and held that post until he was appointed president of the University of Wisconsin in 1887.

Chamberlin is probably best known for his advocacy of the planetesimal theory of the Earth's formation (that the planets evolved from the aggregation of numerous smaller planetlike bodies). In the field of geology, Chamberlin's special interest was glaciers. His 1899 paper "An Attempt to Frame a Working Hypothesis on the Cause of Glacial Periods" tried to show how changes in concentration of carbon dioxide in the air, carbon compounds in rocks, and sea level were related to the growth and retreat of continental glaciers. Chamberlin's theory was not widely accepted, but it was a pioneer effort at showing how all parts of the Earth— atmosphere, hydrosphere, and lithosphere—interact with one another. Chamberlin died in Chicago on November 15, 1928.

James Croll, 1821–1890

Few scholars were as ill prepared to make an impact on climate theory as was James Croll. Born on January 2, 1821, in Cargill, Scotland, as the son of a poor stonemason, Croll could not afford the university training he so badly wanted to have. His formal education ended when he left school at the age of 13.

His poverty did not prevent him from pursuing an ambitious program of self-education, however. He read everything he could get his hands on, especially in the field of science. By the age of 16, he had gained a fair amount of knowledge about heat, light, electricity, magnetism, mechanics, and hydrostatics.

After working as a journeyman millwright and house joiner, he found a position in 1859 as caretaker at Anderson's College and Museum in Glasgow. The new job allowed him to concentrate on the writing of scientific papers. One of these papers dealt with the cause of the ice ages, a topic widely discussed by geologists at the time. Croll argued that the ice ages were caused by changes in the eccentricity of the Earth's orbit around the sun.

In 1867, Croll was offered the position of resident geologist at the newly opened Edinburgh Office of the Geological Survey of Scotland. He served in that position until 1880, when a mild stroke forced him to retire. He died from heart disease in Perth, Scotland, on December 15, 1890. He was the author of three major books, *Climate and Time* (1875), *Discussions on Climate and Cosmology* (1885), and *Stellar Evolution and Its Relation to Geological Time* (1889), and one article of special importance to climate theory, "On the Physical Cause of the Change of Climate during Geological Epochs," which appeared in the August 1864 issue of *Philosophical Magazine*.

Jean Baptiste Joseph Fourier, 1768–1830

Fourier was born in Auxerre, Yonne, France, on March 21, 1768, the son of a poor tailor. His early childhood became even more difficult when he was orphaned at the age of eight. Fourier's first job, a teaching position in his hometown, lasted only three years. In 1798 he joined Napoleon in his military campaign in Egypt. As a reward for his service, Fourier was appointed governor over part of Egypt.

After his return from Egypt in 1801, Fourier concentrated on his mathematical studies. In 1807, he published the research for which he is best known, a theorem that shows how any complex periodic motion can be broken down into some series of simple wave equations.

One of Fourier's special areas of interest was the flow of heat. This topic led him to a consideration of the way heat moves through the atmosphere. He was apparently the first person to compare the Earth's warming with the process that occurs in a greenhouse. He also predicted that human activities might ultimately contribute to this effect and result in a warmer Earth. He summarized a number of speculations about the Earth's greenhouse effect in his 1822 book *Analytic Theory of Heat*. Fourier died in Paris on May 16, 1830, after falling down a flight of stairs.

Albert Gore, Jr., 1948–

Albert Gore, Jr., was born in Washington, D.C., on March 31, 1948. He graduated from Harvard College in 1969 and from Vanderbilt School of Law in 1976. Gore served four terms in the House of Representatives (1976 to 1984) and two terms in the Senate (1984 to 1992). In 1992 he was elected vice-president of the United States.

As a member of Congress, Gore became an expert in the fields of nuclear arms control, the environment, and bioethics. His interest in the problems of climate change go back to the early 1980s. As a young member of the House of Representatives, he held hearings in 1982 and 1984 on the possible impact of human activities on the greenhouse effect. In 1985 he issued a call for an international "Year of the Greenhouse." His book on environmental issues, *Earth in the Balance,* was published in 1992.

James Hansen, 1941–

James Hansen was born in Denison, Iowa, on March 29, 1941. He claims not to have worked very hard as a student, but he managed, nonetheless, to earn a bachelor of arts degree with highest distinction at the University of Iowa in 1963. He then went on to earn a master's degree in astronomy (1965) and a Ph.D. in physics (1967), also at Iowa.

Following graduation, Hansen obtained a postdoctoral fellowship at the Goddard Institute for Space Studies (GISS) in New York City. After two years at GISS, Hansen accepted an appointment at Columbia University as a research associate. He then returned to GISS in 1972. In 1981, he was made head of the institute. Since 1978, Hansen also has been adjunct professor in the Department of Geological Sciences at Columbia.

During his tenure at GISS, Hansen has been credited with taking a poorly known, underfinanced division of NASA and making it into one of the most important centers of global warming research in the world. He also has developed a keen ability to communicate with legislators, the media, and the general public and is, therefore, in constant demand as a spokesman on the topic of climate change.

Hansen has been principal investigator in the *Pioneer* Venus Orbiter Cloud-Photopolarimeter Experiment, the *Galileo* Jupiter Orbiter Photopolarimeter Radiometer Experiment, and the Earth Observing System Interdisciplinary Investigation of Variability of the Earth's Carbon, Energy, and Water Cycles. He has written more than 70 scientific articles and edited or contributed to nearly a dozen books.

Philip Douglas Jones, 1952–

Some of the most important studies of past climate records and predictions of future climate change have come from the Climate Research Unit (CRU) at the University of East Anglia in England. Philip Jones and Tom Wigley have worked together on much of that research.

Jones received his Ph.D. in hydrology from the University of Newcastle upon Tyne in 1977. His dissertation dealt with the modeling of floods. These models were designed to find better methods of regulating river flows. Very soon after his graduation, however, Jones became interested in problems of climate change. In 1981 he began a study of climate change in the Northern Hemisphere from 1851 to 1900, a study funded by the U.S. Department of Energy. Since that time Jones has continued research on climate variability and its relationship to changes in carbon dioxide concentration in the atmosphere.

A number of papers coauthored by Jones have influenced the thinking of scientists and politicians throughout the world. Among these was a 1991 report that the 1980s was the warmest decade in the history of recorded weather measurements.

Philip Jones was born in Redhill, Surrey, on April 22, 1952. He was educated at the Glyn Grammar School in Ewell and earned his bachelor of arts degree in environmental sciences at the University of Lancaster in 1973. A year later he received his master of science degree at the University of Newcastle upon Tyne. Since 1976 he has been senior research associate at the CRU. Jones has coauthored more than 75 scholarly papers and 20 chapters in books on climate change. He also has presented papers at 40 conferences and workshops in 15 nations.

Charles David Keeling, 1928–

If there is any single set of data consistently associated with the topic of global warming, it is probably the "Keeling curve." The Keeling curve is a graph that shows how the concentration of carbon dioxide in the atmosphere has changed over the past 25 years. The data on which this graph is based were collected at the Mauna Loa Observatory in Hawaii by Charles Keeling.

Keeling was born in Scranton, Pennsylvania, on April 20, 1928. He received his bachelor of arts degree in chemistry from the University of Illinois in 1948 and his Ph.D. in chemistry from Northwestern University in 1954. He had just completed a three-year postdoctoral program at the California Institute of Technology when he was selected to head the Mauna Loa project.

Keeling has continued his work in atmospheric chemistry at Scripps Institute of Oceanography to the present day. In 1968 he was appointed professor of oceanography at the institution. Over the past four decades he has received a number of honors and awards, including his selection for the 1980 Second Half Century Award of the American Meteorological Society. The award was given "for his fundamental and far-reaching work on the measurement of atmospheric carbon dioxide which has been the only long-term record of the systematic increase of carbon dioxide in the atmosphere."

Keeling has been a Guggenheim Fellow at the Meteorological Institute of the University of Stockholm (1961 to 1962) and guest professor at both the Zweites Physikalisches Institut of the University of Heidelberg (1969 to 1970) in Germany and the Physikalisches Institut of the University of Bern (1979 to 1980) in Switzerland. He also has served on a number of national and international committees and commissions and has, since 1976, been scientific director of the Central CO_2 Laboratory of the World Meteorological Society.

William W. Kellogg, 1917–

William W. Kellogg was born in New York Mills, New York, on February 14, 1917. He received his bachelor of arts degree in 1939 from Yale University and his master of arts (1942) and Ph.D. in meteorology (1949) from the University of California at Los Angeles (UCLA). He began work in 1947 at the Rand Corporation, where he was head of the Planetary Sciences Department. After 17 years, he left Rand to work at the National Center for Atmospheric Research (NCAR). Between 1949 and 1952, he was also assistant professor in the Institute of Geophysics at UCLA.

Kellogg has been involved in research on climate change for three decades. He joined NCAR in 1964 and was director of its Laboratory of Atmospheric Sciences for ten years. He then served as senior scientist at the laboratory until his retirement in 1987. He was writing about the possible risks of human effects of climate change as early as 1971 and, in 1977, proposed (along with Margaret Mead) the need for a "law of the air." This law, they suggested, would encourage all nations to place specific limits on the amount of carbon dioxide they release into the atmosphere.

Kellogg is perhaps best known in the field of climate change for his research on the possible effects of global warming. As a result of his research in the early 1980s, Kellogg predicted that one potential effect of even small amounts of global warming is a dramatic change in agricultural conditions in some parts of the world.

Syukuro Manabe, 1931–

Syukuro Manabe was born in Shingu-Mura, Japan, on September 21, 1931. He received his bachelor of science (1953), master of science (1955), and doctor of science (1959) degrees from Tokyo University. He became a research meteorologist in the General Circulation Research Section of the U.S. Weather Bureau in 1958. Five years later, he moved to the National Oceanic and Atmospheric Administration's Geophysical Fluid Dynamics Laboratory (GFDL), where he has remained since.

Manabe's special area of expertise is the modeling of factors that may affect global climate. His earliest research dates to the late 1950s, before powerful computers were available. As computer hardware has improved over the past four decades, so have the forecasts produced by the GFDL models. His recent research has concerned the effects of clouds on climate change, projected changes in soil moisture as a result of global warming, and the interaction of atmosphere and oceans in response to increased levels of carbon dioxide.

Milutin Milankovitch, 1879–1958

Milutin Milankovitch was born in Serbia in 1879 into a wealthy family that owned extensive farmlands and vineyards. As a young man, Milankovitch studied agriculture in the expectation of taking over management of the family's lands. But farming was not really his first love, and he eventually entered the Institute of Technology in Vienna to study engineering.

In 1904, Milankovitch earned his doctorate at the institute and immediately went to work building dams and bridges. When offered a job at the University of Belgrade in 1909, he returned to his homeland, where he taught mechanics, astronomy, and theoretical physics.

Throughout his early years, Milankovitch dreamed of finding some great project that would make him famous in the world of science. During an evening of drinking with a poet friend in 1911, Milankovitch made his decision: He would work out a mathematical theory that would explain the temperature of the Earth at different locations on the planet and at different times in its history.

Milankovitch spent the next 30 years of his life on this project. He was interrupted twice in this effort, once by the Balkan War of 1912 and again by the outbreak of World War I in 1914. Eventually he reached his goal, however, by showing how a combination of the Earth's axial precession, its tilt, and its orbital eccentricity result in varying amounts of solar radiation reaching the Earth's surface at different times and places. This combination of factors is now referred to as *the Milankovitch effect.*

Milankovitch's major research is collected in two books, *Mathematical Theory of Heat Phenomena Produced by Solar Radiation* (1920) and *Mathematical Climatology and the Astronomical Theory of Climatic Changes* (1930). He died in 1958 at the age of 79.

Veerabhadran Ramanathan, 1944–

Ramanathan was born in Madras, India, on November 24, 1944. He earned his bachelor's degree from Annamalai University in 1965, his master's at the Indian Institute of Science in Bangalore in 1970, and his doctorate in atmospheric sciences from the State University of New York at Stony Brook in 1974.

After spending a year as a research fellow at NASA's Langley Research Center, he began a long affiliation with the National Center for Atmospheric Research (NCAR), in Boulder, Colorado. He has been leader of the Cloud Climate Interactions Group at NCAR since 1981 and senior scientist there since 1982. He also has held faculty appointments at Colorado State University in Fort Collins since 1985 and the University of Chicago since 1986.

Ramanathan's research has focused on the potential problems posed by greenhouse gases other than carbon dioxide. He has pointed out that trace gases such as CFCs and methane might eventually surpass carbon dioxide in terms of their impact on the greenhouse effect.

Roger Revelle, 1909–1991

Revelle was born in Seattle, Washington, on March 7, 1909, and died in San Diego, California, on July 15, 1991. During his life he was involved in a wide range of research with profound theoretical and practical significance, ranging from the formation of continents and the flow of carbon dioxide through the atmosphere and hydrosphere to the control of human population growth and the improvement of agriculture in Pakistan.

Revelle earned his bachelor of arts degree at Pomona College in 1929 and his Ph.D. at the University of California in 1936. After working as a teaching assistant at Pomona (1929 to 1930) and the University of California (1930 to 1931), he became a research assistant at the Scripps Institute of Oceanography in La Jolla, California. He was affiliated with Scripps on and off, in one capacity or another, until his death.

In the 1950s, Revelle and Hans Suess, also at Scripps, analyzed the movement of carbon dioxide through Earth systems. They found that the oceans take up carbon dioxide much more slowly than scientists had expected. In a 1957 article in the journal *Tellus* they warned that "mankind in spite of itself is conducting a great one-time geophysical experiment," the results of which were almost totally unpredictable.

Stephen Henry Schneider, 1945–

Every important social movement is represented by at least one charismatic, usually outspoken leader. In the civil rights movement of the 1960s, for example, that person was Martin Luther King, Jr. Today, the campaign to make the world more aware of global warming issues is led by at least two brilliant scientists, James Hansen of NASA's Goddard Institute for Space Studies and Stephen H. Schneider at NOAA's National Center for Atmospheric Research (NCAR).

Schneider sometimes seems like a one-man publicity committee on climate change. He has appeared on almost every important news and comment program on television, including the ABC, CBS, and NBC nightly news programs, PBS's *MacNeil-Lehrer Report, Nova, The Today Show, The Tonight Show, Good Morning America, Sunday Morning, NBC Magazine, Nightline, In Search of . . .* , *Who's Who, Universe, 20/20,* The Weather Channel and BBC's *Thames-TV* and *Horizon.*

He also has been interviewed on Australian, Austrian, Canadian, British, Italian, Chilean, and Japanese national commercial radio and television programs. In addition, he has served as script consultant or scientific advisor for a number of popular radio and television programs, including *Nova*'s "The Climate Crisis" and "Goddess of the Earth," "A Change in the Weather" (KPIX, San Francisco), "Planet Earth Project" (WQED, Pittsburgh), and "Nuclear Winter: Ethics of Dispute" (KRMA, Denver).

Schneider's popular books and other writings on climate change—especially *The Genesis Strategy, Global Warming,* and *The Coevolution of Climate and Life*—provide some of the most readable discussions of global warming issues available anywhere.

Stephen Schneider was born in New York City on February 11, 1945. He received his bachelor of science (1966), master of science (1967), and Ph.D. in mechanical engineering (1971) degrees from Columbia University. His first professional assignment was with James Hansen's research group at NASA's Goddard Institute for Space Studies from 1971 to 1972. In 1972 he joined NCAR, where he has been employed ever since. He is now head of the Interdisciplinary Climate Systems Section at NCAR.

Schneider has received a number of professional and public awards, including the American Association for the Advancement of Science's 1991 AAAS-Westinghouse Award for Public Understanding of Science and Technology and a 1992 MacArthur Foundation Fellowship.

James Gustave Speth, 1942–

Research and education about climate change have long involved both governmental agencies and nongovernmental organizations. One of the most effective of the latter has been the World Resources Institute (WRI), an independent research and policy center in Washington, D.C., funded by private foundations, the United Nations, governmental agencies, corporations, and individuals. WRI attempts to help government, private industry, and individuals find ways to attain economic growth without degrading the natural environment. The cofounder and current president of WRI is Gus Speth.

James Gustave Speth was born in Orangeburg, South Carolina, in 1942, where he attended public schools. After graduating from Yale University in 1964, Speth attended Balliol College, Oxford University, on a Rhodes scholarship for two years. Upon his return to the United States, he entered Yale Law School, from which he received his degree in 1969.

Speth's special interest has long been environmental law. In 1970, he helped found the Natural Resources Defense Council (NRDC), an organization of which he became senior attorney. In 1976 he received the National Wildlife Foundation's Resources Defense Award for his work at NRDC.

From 1977 to 1982, Speth served on the President's Council on Environmental Quality (CEQ), the last two years as chair of the agency. In 1982, he left the council and helped form the WRI. For a number of years Speth has been especially outspoken on issues of climate change. As early as 1981, for example, CEQ issued a report warning about the potential risks posed by the continued release of high levels of carbon dioxide as a result of human activities. In 1987, Speth proposed that Congress consider comprehensive legislation on methods for dealing with global warming. His suggestions were later incorporated in the 1988 Global Climate Protection Act.

Maurice Strong, 1929–

Maurice Strong served as secretary-general of the United Nations Conference on Environment and Development held in June 1992 in Rio de Janeiro, Brazil. Strong is a rare individual who brings together two normally contrasting backgrounds. He is both a very successful, very rich businessman who made his fortune in mining and a passionate environmentalist. One colleague has noted that "there is this contradiction which most of us don't have. We're either poor and committed or rich and carefree. Somehow [Strong] manages to be rich and committed."

Strong is the classic self-made millionaire. He was born in Oak Lake, Manitoba, on April 29, 1929. His father, a telegraph operator for the Canadian-Pacific Railway, lost his job in the Great Depression. Young Maurice began his education in the public schools of Oak Lake. But, discouraged by the poverty and boredom of his life, he ran away from home at the age of 14.

One of Strong's first jobs was as a security officer at the newly created United Nations. He then went on to work as a fur trader in Alaska and to operate a trading post in northern Canada for the Hudson Bay Company. Eventually he became involved in oil exploration, with rapid success. He made his first million by the age of 23.

His commitment to human and environmental concerns goes back at least 30 years. In 1966 he gave up his job with a giant Canadian power company to become head of Canada's overseas aid program. In 1970 he was appointed secretary-general of the 1972 United Nations Conference on the Human Environment, held in Stockholm, Sweden.

Strong has had his share of critics in the past. Some environmentalists complain about his close ties to business. They say that some of his own companies' projects have contributed to the very environmental problems against which he claims to fight. For Strong, however, the tie between environmental issues and commercial development is absolutely essential. Echoing the theme of the 1992 Rio conference, he asked, "How can you envisage a change of course to sustainable development without making the business community peers and allies? It simply can't be done."

Margaret Thatcher, 1925–

In the fall of 1988, Great Britain's first woman prime minister, Margaret Thatcher, gave an important speech before the Royal Society. In her speech, Thatcher expressed the view that global warming was a critical issue to which Britain and the whole world must turn its attention. "Protecting the balance of nature," she said, "is one of the great challenges of the twentieth century."

Thatcher's views were particularly striking because, according to author John Gribbins, she had had "an abysmal record on environmental issues" for most of her tenure in office. After 1988, however, the British government began to provide enthusiastic support for research on climate change.

Margaret Thatcher was born in Lincolnshire, Great Britain, on October 13, 1925. Her education began at Kesteven and Grantham Girls' schools in Lincolnshire. Then she won a scholarship to Somerville College, Oxford, where she majored in chemistry. After graduation she took a job as research chemist with a company that manufactures celluloid.

Thatcher was first elected to the House of Commons in 1959. She served her party in a number of posts over the next 15 years and, in 1975, was chosen leader of the Conservative Party. In the general election of 1979, the Labour government fell and Thatcher became prime minister. She served in that office until resigning in favor of John Majors in 1990.

John Tyndall, 1820–1893

Tyndall was born in Leighlinbridge, Carlow, in what is now the Republic of Ireland, on August 2, 1820. His most illustrious ancestor was William Tyndall, a translator of the Bible who was burned at the stake as a heretic in 1536.

After working as a civil servant and a railway engineer, Tyndall enrolled at the University of Marburg, from which he received his doctoral degree in 1851. Three years later he was appointed professor of natural philosophy at the Royal Institution. He died of an accidental overdose of sleeping medicine in Hindhead, Surrey, on December 4, 1893.

An area of special interest to Tyndall was the subject of heat. In one experiment on heat, he measured the absorption of infrared radiation by water vapor and carbon dioxide. Based on this research, he concluded that the ice ages may have occurred as a result of a decrease in the concentration of carbon dioxide in the atmosphere, an effect caused by some unknown factor. He reported this idea in the article "On Radiation throughout the Earth's Atmosphere" in the *Philosophical Magazine* in 1863. Tyndall's theory appears to be the first one that connected carbon dioxide in the atmosphere with massive climate change such as the ice ages.

Harold Urey, 1893–1981

Harold Urey was born in Walkerton, Indiana, on April 29, 1893. He earned a bachelor of science degree in zoology at the University of Montana in 1917 and a Ph.D. in chemistry at the University of California in 1923. After a year of postgraduate study at the Bohr Institute in Copenhagen, Denmark, Urey accepted an appointment at Johns Hopkins University (1924–1929) and then moved to Columbia University, where he remained until 1945. In that year, he was appointed Martin A. Ryerson Distinguished Service Professor at the University of Chicago. He died on January 5, 1981, in La Jolla, California.

Urey is perhaps most famous for his discovery in 1931 of deuterium, the heavy isotope of hydrogen. For this accomplishment, he was awarded the 1934 Nobel Prize in chemistry. As a result of his work with isotopes, Urey also became interested in the subject of climate change. He found that the ratio of two different isotopes of oxygen in aquatic fossils could be used to estimate the temperature of ocean water in which the animals had originally lived. This discovery has become an important technique for determining the Earth's temperature at various points in its history.

Richard T. Wetherald, 1936–

Richard T. Wetherald was born in Plainfield, New Jersey, on March 28, 1936. He received his bachelor of science and master of science degrees at the University of Michigan in 1962 and 1963, respectively. After a year with the Westinghouse Electric Company, Wetherald joined NOAA's Geophysical Fluid Dynamics Laboratory as a research meteorologist. He is author or coauthor of 20 papers on climate modeling.

Tom Michael Lampe Wigley, 1940–

Tom Wigley was born in Adelaide, South Australia, on January 18, 1940. He earned his bachelor of science degree at the University of Adelaide in 1959, majoring in pure math, applied math, and physics. He then took a course at the Commonwealth Bureau of Meteorology to become a meteorologist before returning to the University of Adelaide for his doctoral work. In 1967, he was awarded a Ph.D. in mathematical physics, specializing in problems of physical dynamics and fluid mechanics.

From 1968 to 1975, Wigley taught applied mathematics, statistics, air pollution, and meteorology at the University of Waterloo in Ontario, Canada. He left Waterloo in 1975 to begin work in climatology at the Climate Research Unit (CRU) at Great Britain's University of East Anglia. When H. H. Lamb retired as director of the unit in 1978, Wigley was appointed to replace him.

Under Wigley's direction, the CRU has become one of the world's leading centers for climate research. It maintains one of the four major general circulation models in the world. Wigley's own research at CRU has focused on the impact of climate on agriculture and water resources; climate, sea level, and carbon cycle modeling; and paleoclimatology. He has authored or coauthored more than 150 papers and reports on climate and has served as consultant to more than a dozen international agencies.

Timothy E. Wirth, 1939–

Timothy E. Wirth was born in Santa Fe, New Mexico, on September 22, 1939. He earned a bachelor's degree in history (1961) and a master's degree in education (1964) from Harvard University and a doctorate in education (1973) from Stanford University. He was elected to six terms in the House of Representatives from Colorado (1974–1986) and to one term in the Senate. He did not run for reelection to the Senate in 1992.

In the Senate, Wirth became a member of the Senate Energy and Natural Resources Committee. He quickly developed an interest in environmental problems, concerning himself especially with the issue of global warming. Wirth has held a series of hearings at which witnesses presented evidence of warming trends and has raised the consciousness of both legislators and the general public about future risks resulting from greenhouse gas emissions.

4

Facts, Data, and Opinion

THIS CHAPTER BRINGS TOGETHER some of the fundamental information available about global warming as well as the interpretations of this information by different individuals and organizations. The chapter is divided into four major subsections: "Statistical Data," "Predictions about Global Warming," "Possible Impacts of Global Warming," and "Recommendations for Action."

The debate over global warming begins with certain factual information about which there tends to be general agreement among scientists. These data include the amount of fossil fuels burned each year, the concentration of carbon dioxide and other trace gases in the atmosphere, and changes that have taken place in the average annual global temperature in history.

Authorities differ as to how the data about global warming should be interpreted. Some argue that these data show that global warming has already begun. Others doubt that the evidence for warming exists although they do believe that the conditions that will produce warming are already present. Still others are not convinced that existing data are sufficient to conclude that warming is certain or even likely.

Those who believe global warming is now, or soon will be, a reality have tried to speculate about the possible impacts of a warmer globe. They have tried to predict effects on ocean levels, agricultural practices, forestry, water resources, urban life, food production, and other aspects of the environment and human society.

Faced with the information that exists about global warming and the remaining uncertainties as to what the future holds, scientists, politicians, and others have made a wide variety of suggestions as to how the world should deal with the risk of possible global warming. Some of these suggestions are "technological fixes," which are programs for counteracting existing greenhouse gas emissions by various scientific means. Others are social and political programs for dealing with increasing greenhouse gas emissions. These include a whole range of ideas from doing nothing at all to making immediate and radical changes in the social, economic, political, and psychological aspects of human civilization.

Statistical Data

Underlying the discussion of global warming are some basic facts about which most people can agree. It would be nice to believe that some cold, hard facts exist on any scientific topic, facts on which everyone can agree without dispute. Unfortunately, that is almost never the case. Instead there is a range of statistical data about which people have a greater or lesser degree of confidence. The sections that follow contain data about which experts tend to have a high degree of confidence, although there may be some dispute about any of the charts in terms of the validity of the way the data were collected.

Consumption of Fossil Fuels

Global warming, if it occurs, results from an increase in the concentration of carbon dioxide and other greenhouse gases in the atmosphere. Carbon dioxide enters the atmosphere primarily as a result of the burning of coal, oil, and natural gas. Tables 4.1 and 4.2 show how the release of carbon dioxide as a result of the use of these fuels has changed in the world over the past 130 years. The release of carbon dioxide from cement production and gas flaring (the burning of waste natural gas) is also listed.

The tables show that coal was by far the nation's most popular fuel until the 1920s. Then the use of petroleum and natural gas began to increase rapidly. By 1968, liquid fuels (primarily petroleum) had bypassed solid fuels (primarily coal and lignite, a cheap grade of coal) as the fuels of choice in the United States.

Table 4.1.

Estimated Emissions from Various Sources: Historical Records (1860–1953)*

Year	Coal	Lignite	Crude Petroleum	Natural Gas	Cement Production	Total
			Estimated Emissions from the Burning of			
1860	91.5	1.7	0.1	0.0	NA (Not applicable)	93.3
1861	96.7	1.8	0.2	0.0	NA	98.7
1862	96.1	2.0	0.3	0.0	NA	98.4
1863	103.5	2.1	0.3	0.0	NA	106.0
1864	112.5	2.4	0.2	0.0	NA	115.1
1865	119.0	2.6	0.3	0.0	NA	121.9
1866	125.7	2.6	0.4	0.0	NA	128.7
1867	134.7	2.8	0.4	0.0	NA	137.9
1868	133.3	3.0	0.4	0.0	NA	136.7
1869	138.1	3.2	0.5	0.0	NA	141.8
1870	141.0	3.4	0.6	0.0	NA	145.0
1871	157.4	3.9	0.6	0.0	NA	161.9
1872	170.0	4.3	0.7	0.0	NA	175.9
1873	182.5	4.8	1.2	0.0	NA	188.4
1874	177.3	5.3	1.2	0.0	NA	183.8
1875	182.8	5.3	1.1	0.0	NA	189.2
1876	184.9	5.5	1.2	0.0	NA	191.6
1877	188.8	5.5	1.7	0.0	NA	196.0
1878	188.9	5.7	1.9	0.0	NA	196.5
1879	199.1	6.0	2.5	0.0	NA	207.6
1880	217.5	6.4	3.2	0.0	NA	227.1
1881	234.0	6.8	3.4	0.0	NA	244.4
1882	251.3	7.1	3.8	0.3	NA	262.5
1883	268.8	7.6	3.2	0.4	NA	280.0
1884	269.9	7.8	3.8	0.6	NA	282.1
1885	263.8	8.0	3.8	0.8	NA	276.4
1886	264.4	8.3	5.0	1.0	NA	278.7
1887	282.8	8.6	5.0	1.3	NA	297.7
1888	305.8	9.2	5.5	1.5	NA	321.9
1889	310.5	9.8	6.5	1.7	NA	328.5
1890	328.9	10.8	8.1	2.0	NA	349.7
1891	342.2	11.5	9.6	2.0	NA	365.4
1892	345.5	11.7	9.4	2.1	NA	368.7
1893	337.6	12.2	9.7	2.2	NA	361.6
1894	352.9	12.5	9.4	2.3	NA	377.1
1895	371.7	13.6	10.9	2.4	NA	398.6
1896	382.4	14.5	12.1	2.7	NA	411.6
1897	400.0	15.8	12.9	3.0	NA	431.5
1898	421.5	16.7	13.2	3.2	NA	454.6
1899	462.2	17.7	13.8	3.5	NA	497.3
1900	485.6	19.8	15.8	3.7	NA	524.9
1901	497.2	21.3	17.7	4.1	NA	540.3
1902	508.5	20.9	19.1	4.4	NA	552.9
1903	559.5	21.7	20.5	4.7	NA	606.4
1904	563.0	22.6	23.0	4.9	NA	613.4
1905	594.2	24.1	22.7	5.6	NA	646.6
1906	641.6	25.8	22.5	6.1	NA	696.1
1907	708.7	28.3	27.8	6.5	NA	771.2
1908	670.9	29.3	30.1	6.3	NA	736.6

(Table 4.1 *continued*)

			Estimated Emissions from the Burning of			
Year	Coal	Lignite	Crude Petroleum	Natural Gas	Cement Production	Total
1909	700.2	29.8	31.5	7.5	NA	769.0
1910	732.2	30.0	34.5	8.0	NA	804.8
1911	746.2	31.2	36.3	8.1	NA	821.8
1912	785.7	34.5	37.1	8.9	NA	866.2
1913	842.9	35.7	41.3	9.2	NA	929.0
1914	752.7	33.5	43.0	9.3	NA	838.4
1915	741.2	34.3	45.4	9.9	NA	830.8
1916	798.4	36.4	48.1	11.8	NA	894.8
1917	842.2	37.4	53.2	12.5	NA	945.3
1918	828.1	39.2	53.4	11.3	NA	932.0
1919	720.7	36.9	59.6	11.7	NA	828.9
1920	826.4	43.6	76.3	12.6	NA	958.9
1921	688.3	46.6	82.6	10.4	NA	828.0
1922	735.8	50.6	92.2	12.0	NA	890.6
1923	835.4	45.3	108.8	15.8	NA	1005.3
1924	824.0	47.9	108.7	17.9	NA	998.5
1925	321.3	51.8	114.4	18.9	NA	1006.4
1926	815.7	52.0	117.5	20.8	NA	1006.4
1927	883.7	56.2	134.7	22.8	NA	1097.5
1928	863.5	60.8	141.6	24.9	NA	1090.0
1929	918.3	64.4	158.9	30.3	NA	1171.9
1930	843.3	54.6	151.2	28.4	NA	1077.5
1931	744.6	50.3	145.7	27.6	NA	968.2
1932	662.0	47.1	139.0	25.7	NA	873.8
1933	693.0	48.3	151.6	25.9	NA	918.8
1934	754.0	53.0	160.1	29.4	NA	996.5
1935	770.3	57.0	174.3	30.1	NA	1031.7
1936	866.3	56.8	188.4	34.9	NA	1146.4
1937	902.3	69.8	215.3	38.8	NA	1226.2
1938	841.3	73.1	209.9	37.0	NA	1161.4
1939	892.6	80.9	219.6	39.8	NA	1232.9
1940	944.6	88.4	224.5	42.9	NA	1300.4
1941	985.4	91.7	214.6	45.4	NA	1337.1
1942	990.3	92.5	202.2	49.4	NA	1334.4
1943	991.0	96.4	221.5	55.1	NA	1364.0
1944	955.6	87.5	249.2	59.8	NA	1352.2
1945	823.3	51.9	263.0	65.5	NA	1203.6
1946	842.7	70.9	289.1	67.8	NA	1270.5
1947	952.2	75.9	318.4	75.0	NA	1421.5
1948	990.3	82.9	359.9	84.5	NA	1517.5
1949	930.0	92.0	358.4	89.2	15.8	1485.4
1950	1007.6	99.9	402.2	103.2	18.3	1631.2
1951	1056.1	109.9	479.0	115.0	20.6	1780.6
1952	1038.8	117.2	504.0	124.0	22.1	1806.1
1953	1039.5	124.7	506.0	138.0	24.6	1832.8

*In million metric tons of carbon equivalents.
Source: Various references as cited in Thomas A. Boden, Robert J. Sepanski, and Frederick W. Stoss, eds., *Trends '91: A Compendium of Data on Global Change* (Oak Ridge, TN: Carbon Dioxide Information Analysis Center, December 1991), p. 385.

Table 4.2.

Estimated Emissions from Various Sources: Modern Records (1950–1989)*

Year	Total	Solid	Liquid	Gas	Cement Production	Gas Flaring	Per Capita Emissions**
1950	1638	1077	423	97	18	23	0.65
1951	1775	1137	479	115	20	24	0.69
1952	1803	1127	504	124	22	26	0.69
1953	1848	1132	533	131	24	27	0.70
1954	1871	1123	557	138	27	27	0.69
1955	2050	1215	625	150	30	31	0.74
1956	2185	1281	679	161	32	32	0.78
1957	2278	1317	714	178	34	35	0.80
1958	2338	1344	732	192	35	35	0.80
1959	2471	1390	790	214	40	36	0.83
1960	2586	1419	850	235	43	39	0.86
1961	2602	1356	905	254	45	42	0.85
1962	2708	1358	981	277	49	44	0.86
1963	2855	1404	1053	300	51	47	0.89
1964	3016	1442	1138	328	57	51	0.92
1965	3154	1468	1221	351	59	55	0.95
1966	3314	1485	1325	380	63	60	0.97
1967	3420	1455	1424	410	65	66	0.98
1968	3596	1456	1552	445	70	73	1.01
1969	3809	1494	1674	487	74	80	1.05
1970	4091	1571	1838	516	78	87	1.11
1971	4242	1571	1946	554	84	88	1.12
1972	4409	1587	2056	583	89	94	1.15
1973	4648	1594	2240	608	95	110	1.18
1974	4656	1591	2244	618	96	107	1.16
1975	4629	1686	2131	623	95	93	1.13
1976	4895	1723	2313	647	103	109	1.18
1977	5034	1786	2390	646	108	104	1.19
1978	5082	1802	2383	674	116	107	1.18
1979	5366	1899	2535	714	119	100	1.23
1980	5264	1921	2409	725	120	89	1.18
1981	5129	1930	2272	735	121	72	1.13
1982	5094	1993	2178	734	121	69	1.11
1983	5085	1998	2163	736	125	63	1.09
1984	5243	2088	2186	785	128	57	1.10
1985	5369	2196	2169	819	131	55	1.11
1986	5551	2253	2274	836	136	52	1.12
1987	5661	2310	2285	876	142	48	1.13
1988	5897	2388	2388	916	150	55	1.15
1989	5967	2392	2419	946	152	56	1.15

*In million metric tons of carbon equivalents.

**In metric tons of carbon equivalent.

Source: Various references as cited in Thomas A. Boden, Robert J. Sepanski, and Frederick W. Stoss, eds., *Trends '91: A Compendium of Data on Global Change* (Oak Ridge, TN: Carbon Dioxide Information Analysis Center, December 1991), p. 389.

That pattern changed once again in the 1980s, as solid fuels once more became more popular. Since 1983, the total emissions from solid and liquid fuels have remained approximately equal.

Production of CFCs

It is becoming increasingly clear that gases other than carbon dioxide contribute to global warming. Indeed, it appears the contribution of all other greenhouse gases may be at least equal to that of carbon dioxide. Unfortunately, no data comparable to fossil fuel combustion exist for estimating the amount of methane, nitrogen oxides, and other greenhouse gases. The only exception to that statement concerns the chlorofluorocarbons (CFCs), for which production and release data exist over a 60-year period. That period is shorter than it is for fossil fuel use but is sufficient to show trends for this important greenhouse gas.

Figure 4.1 shows the trends in estimated release of CFC-11 and CFC-12, the two most commonly used CFCs. The graph

Figure 4.1.
Estimated Release of CFCs, 1931–1989

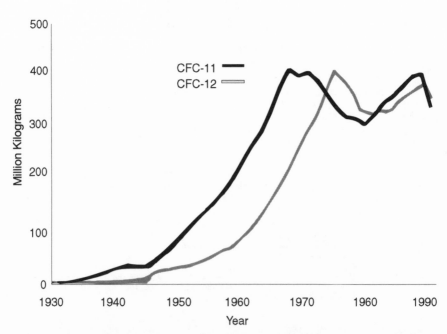

Source: Fluorocarbon Program Panel of the Chemical Manufacturers Association and other sources as reported in Thomas A. Boden, Robert J. Sepanski, and Frederick W. Stoss, eds., *Trends '91: A Compendium of Data on Global Change* (Oak Ridge, TN: Carbon Dioxide Information Analysis Center, December 1991), pp. 373, 377.

shows that emissions of these two compounds peaked in the mid-1970s and have now leveled off at about 90 percent of their maximum level. Both compounds are scheduled to be removed from production in the late 1990s as a result of the Montreal Protocol on ozone depletion.

Carbon Dioxide Concentrations (Keeling Curve)

The single piece of evidence that may be most influential in convincing scientists that at least the potential for global warming is now present is the so-called Keeling curve. This curve shows the concentration of carbon dioxide as measured at the top of Mauna Loa in Hawaii over the past 35 years. Figure 4.2 shows that the concentration of carbon dioxide (in parts per million by volume) in this location has increased regularly every year since monitoring

Figure 4.2.
Monthly Atmospheric Concentrations of Carbon Dioxide
at the Mauna Loa Monitoring Station

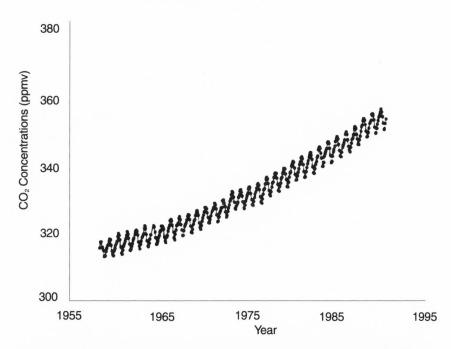

Source: Mauna Loa Observatory Data as collected by Charles D. Keeling et al. and reported in Thomas A. Boden, Robert J. Sepanski, and Frederick W. Stoss, eds., *Trends '91: A Compendium of Data on Global Change* (Oak Ridge, TN: Carbon Dioxide Information Analysis Center, December 1991), pp. 13–15.

began in 1955. The annual fluctuations represent changes that occur as a result of cyclic growing seasons.

Methane Concentrations

Data regarding the concentration of methane in the atmosphere have been available for a shorter time than the carbon dioxide data shown in Figure 4.2. These data, however, show similar trends to those shown by the carbon dioxide data. Figure 4.3 shows the average concentration of methane (in parts per billion by volume) in the atmosphere, as reported by a number of monitoring stations in various locations around the planet. Notice that the overall trend is upward although the seasonal fluctuations observable with carbon dioxide do not hold true for methane.

Figure 4.3.
Monthly Global Concentrations of Atmospheric Methane
(CH₄) Concentrations (parts per billion by volume)

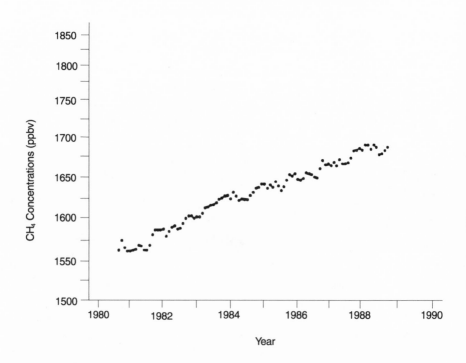

Source: Data collected by M. Aslam K. Khalil et al. at the Oregon Graduate Institute of Science and Technology as reported in Thomas A. Boden, Robert J. Sepanski, and Frederick W. Stoss, eds., *Trends '91: A Compendium of Data on Global Change* (Oak Ridge, TN: Carbon Dioxide Information Analysis Center, December 1991), pp. 228–231.

CFCs and N₂O Concentrations

Patterns of growth for CFCs and nitrous oxide (N₂O) concentrations are similar to those for carbon dioxide and methane. Figure 4.4 shows changes in the atmospheric concentration of the two most commonly used CFCs—CFC-11 and CFC-12—and of nitrous oxide. Figure 4.4 shows a steady increase in the amount of

Figure 4.4.
Monthly Atmospheric Concentrations of CFCs and Nitrous Oxide

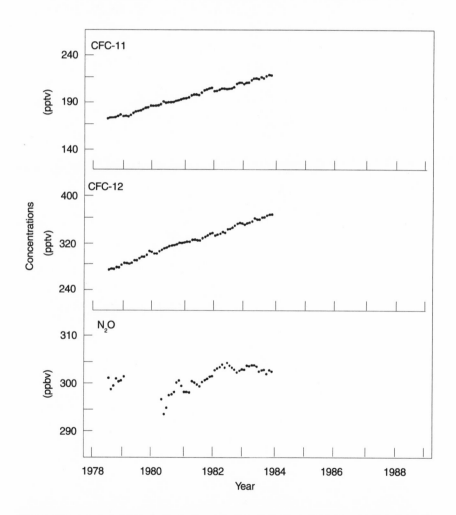

Source: Data collected at Adrigole, Bantry Bay, Ireland, by F. N. Alyea, D. M. Cunnold, R. G. Prinn. R. Rasmussen, S. Crawford, P. Simmonds, P. Fraser, and R. Rosen as reported in Thomas A. Boden, J. Sepanski, and Frederick W. Stoss, eds., *Trends '91: A Compendium of Data on Global Change* Ridge, TN: Carbon Dioxide Information Analysis Center, December 1991), pp. 340–345.

CFCs found in the atmosphere since monitoring began in 1978 and a less regular, but still increasing, pattern for nitrous oxide.

Historical Carbon Dioxide Trends (Vostok)

For a number of years, scientists have been studying an ice core 2,083 meters (6,832 feet) long, taken by the Soviet Antarctic Expedition at Vostok in the East Antarctica. They have been able to determine the concentration of carbon dioxide (in parts per million by volume) in sections of the core dating back about 160,000 years. Figure 4.5 shows the concentration of carbon dioxide found in ice cores going back about 164,000 years. The results of this

Figure 4.5.
Carbon Dioxide Concentration in Vostok Core for Various Ice Ages

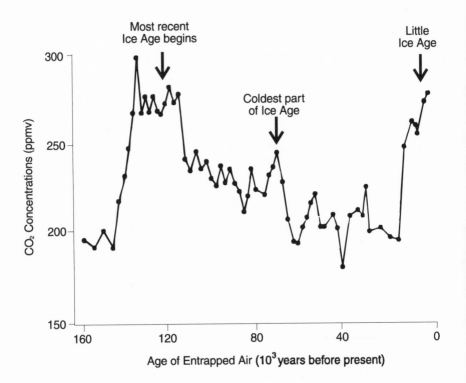

Age of Entrapped Air (10^3 years before present)

Source: Data collected at Vostok Station, Antarctica, by J. M. Barnola, D. Raynaud, C. Lorius, and Y. S. Korotkevich as reported in Thomas A. Boden, Robert J. Sepanski, and Frederick W. Stoss, eds., *Trends '91: A Compendium of Data on Global Change* (Oak Ridge, TN: Carbon Dioxide Information Analysis Center, December 1991), pp. 4–7.

research show that carbon dioxide levels in the atmosphere are related to climate changes, such as the ice ages about which we already know. The conclusion seems to be that varying the concentration of carbon dioxide in the atmosphere may indeed have some effect on the Earth's overall climate patterns.

Temperature Trends, Twentieth Century

Temperature changes over the past century have been studied by a number of researchers. Figure 4.6 shows temperature anomalies over the past century. The anomaly for each year is the deviation of that temperature for that year from the average temperature for the period 1940–1960. The data have been corrected for the disturbing effect of El Niño and Southern Oscillation (ENSO) events. Figure 4.6 shows that the observed temperature variations are highly irregular but appear to show a gradual increase over the past century.

Historical Sea-Level Trends

As the Earth's annual average temperature rises, parts of the ice caps and glaciers will begin to melt. Water thus produced will travel to the ocean and cause sea levels to rise. However, an even more important effect is the expansion that ocean water will undergo as it is warmed. Figure 4.7 plots the estimated annual changes in sea level over the past century alongside the five-year mean changes over the same period. The five-year mean takes

Figure 4.6.
Global Annual Temperature Anomalies, 1854–1990

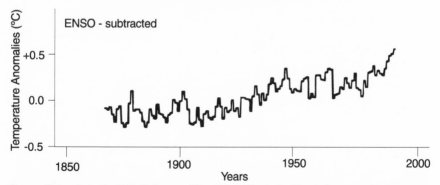

Source: Data collected by Philip D. Jones and Tom M. L. Wigley at the Climatic Research Unit, University of East Anglia, as reported in Thomas A. Boden, Robert J. Sepanski, and Frederick W. Stoss, eds., *Trends '91: A Compendium of Data on Global Change* (Oak Ridge, TN: Carbon Dioxide Information Analysis Center, December 1991), pp. 512–517.

Figure 4.7.
Estimated Sea-Level Trends, 1880–1980

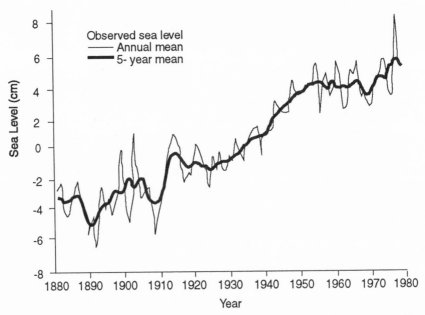

Source: V. Gornitz, S. Lebedeff, and J. Hansen, "Global Sea Level Trend in the Past Century," *Science*, March 26, 1982.

into account annual fluctuations. The two curves thus provide a more accurate indication of changes than either set of data alone. The graph suggests that sea levels appear to follow the same general trend as that of temperature during the same period (refer to Figure 4.6 for temperature changes).

The Carbon Cycle

To understand the impact of carbon dioxide on climate, it is necessary to know what happens to carbon dioxide in the Earth's atmosphere, hydrosphere, and lithosphere. Figure 4.8 shows the amount of carbon stored in each of these three reservoirs and the exchanges that take place among them. Carbon dioxide is released to the atmosphere primarily as a result of combustion of fuels, respiration by animals, and decomposition of dead organisms. It returns to the form of chemical compounds primarily by means of photosynthesis, in which plants convert carbon dioxide and water to complex organic compounds.

Figure 4.8.
The Global Carbon Cycle

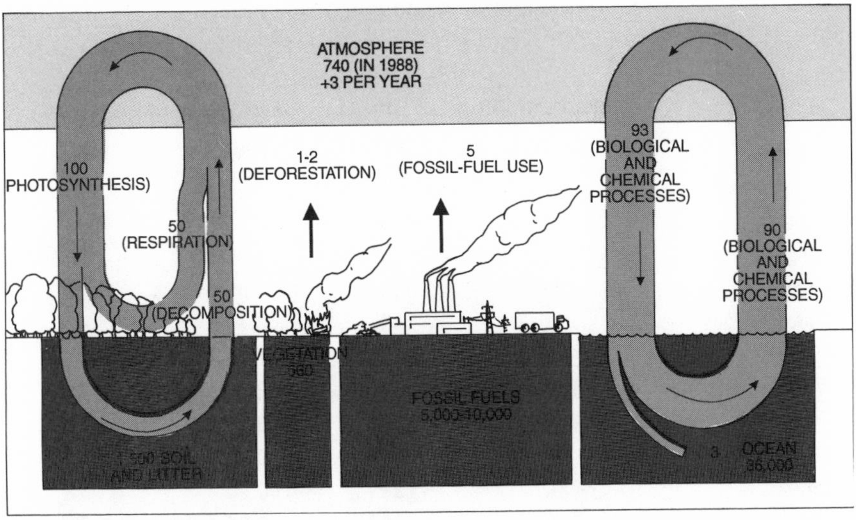

Source: Congress, Office of Technology Assessment, *Changing by Degrees: Steps To Reduce Green-house Gases,* OTA-O-482 (Washington, DC: Government Printing Office, February 1991), p. 59.

Production of Carbon Dioxide

Assuming that it might be desirable to reduce the amount of carbon dioxide and other greenhouse gases released to the atmosphere, a first step would be to find out where those gases originate on the Earth. Table 4.3 and Figures 4.9 and 4.10 provide some of this information. Table 4.3 shows the amount of carbon dioxide (calculated as carbon) produced per person in various countries of the world. Notice that the United States, Canada, Australia, the former Soviet Union, and Saudi Arabia are the largest single producers of carbon dioxide per capita in the world.

Figure 4.9 shows the contribution made by each greenhouse gas to radiative forcing. This diagram shows that carbon dioxide accounts for just over half of the total greenhouse effect and that methane and the CFCs are the next most important gases in this respect. Figure 4.10 shows how various sectors of the U.S. economy contribute to carbon dioxide emission. The three major sectors, buildings, transportation, and industry, each contribute about equally to the total release of carbon dioxide.

Table 4.3.
Per Capita Release of Carbon Dioxide from Various
Countries in the World (1960 and 1987)

Country	Carbon (millions of tons) 1960	1987	Carbon per capita (tons) 1960	1987
United States	791	1,224	4.38	5.03
Canada	52	110	2.89	4.24
Australia	24	65	1.85	3.68
Soviet Union	396	1,035	1.85	3.68
Saudi Arabia	1	45	0.18	3.60
Poland	55	128	1.86	3.38
West Germany	149	182	2.68	2.98
United Kingdom	161	156	3.05	2.73
Japan	64	251	0.69	2.12
Italy	30	102	0.60	1.78
France	75	95	1.64	1.70
South Korea	3	44	0.14	1.14
Mexico	15	80	0.39	0.96
China	215	594	0.33	0.56
Egypt	4	21	0.17	0.41
Brazil	13	53	0.17	0.38
India	33	151	0.08	0.19
Indonesia	6	28	0.06	0.16
Nigeria	1	9	0.02	0.09
Zaire	1	1	0.04	0.03
WORLD	2,547	5,599	0.82	1.08

Source: Christopher Flavin, *Slowing Global Warming: A Worldwide Strategy* (Washington, DC: Worldwatch Institute, 1989), p. 26 as taken from various sources.

Characteristics of Greenhouse Gases

Various greenhouse gases have different characteristics and different potential effects on global warming. In Table 4.4, the category Global Warming Potential (GWP) is defined as the ratio of the amount of warming from a pound of a greenhouse gas to the warming from a pound of carbon dioxide over a certain period of 20, 100, or 500 years. The compounds designated as HCFC-22, HCFC-123, and so on, are hydrochlorofluorocarbons, compounds that have been designed to replace chlorofluorocarbons because they tend to have less harmful effects on ozone in

Figure 4.9.
Relative Contribution of Each Greenhouse Gas
to Global Warming in the 1980s

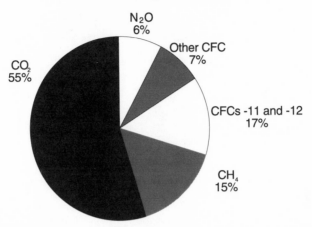

Source: Intergovernmental Panel on Climate Change, *Scientific Assessment of Climate Change,* Summary Report, World Meteorological Organization/U.N. Environment Program (Cambridge, England: Cambridge University Press, 1990).

the atmosphere. Notice that they will still make major contributions to global warming, however.

Emissions Due to Deforestation

Many observers are concerned about the effect of deforestation in producing global warming. They fear that the destruction of tropical rain forests removes a critical sink for carbon dioxide in the atmosphere, thus worsening the warming effects of that gas. Table 4.5 summarizes the estimated carbon emissions contribution of various nations to global warming as a result of deforestation. The table points out that the major contributors to global warming as a result of deforestation are Third World countries with large tropical forests, such as Brazil, Indonesia, and Colombia.

Predictions about Global Warming

An Early Warning about Warming

One of the earliest scientific warnings about the possibility of an enhanced greenhouse effect as the result of human activities appeared in the scientific journal *Tellus* in 1957. In this article,

Figure 4.10.
Emission of Carbon Dioxide in the United States
by Various Sectors of the Economy

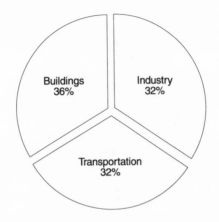

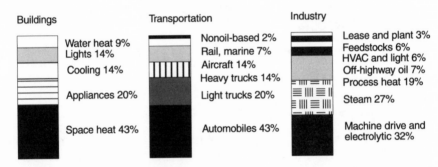

Source: Revised from U.S. Congress, Office of Technology Assessment, *Changing by Degrees: Steps to Reduce Greenhouse Gases*, OTA-O-482 (Washington, DC: Government Printing Office, February 1991), p. 8.

"Carbon Dioxide Exchange between Atmosphere and Ocean and the Question of an Increase of Atmospheric CO_2 During the Past Decades," Roger Revelle and Hans Suess ask how anthropogenic carbon dioxide could have significant effects on the Earth's average annual temperature and, therefore, its future climate.

> During the next few decades the rate of combustion of fossil fuels will continue to increase, if the fuel and power requirements of our worldwide industrial civilization continue to rise exponentially, and if these needs are met only to a limited degree by development of atomic power. Estimates by the UN (1955) indicate that during the first decade of the 21st century fossil fuel combustion could produce an amount of carbon dioxide equal to 20% of that now in the atmosphere.

Table 4.4.
Some Important Characteristics of Greenhouse Gases

Gas	Assumed concentration in 1880[1]	Concentration in 1990[2]	Projected concentration in 2030[3]	Annual growth rate as of 1990 (percent/year)	Contribution to warming 1880–1980[1]	Lifetime (years)	GWP[2]
CO_2	260-290*	353*	440-450*	0.5	66%	120	1
CH_4	1.2*	1.72*	2.5-2.6*	0.9	15%	10	21
N_2O	290**	310**	340**	0.25	3%	150	290
CFC-11	0	0.28**	0.5**	4.0	4%	60	3,700
CFC-12	0	0.48**	1.0-1.1**	4.0	5%	130	7,600
Others	. . . . .NO ESTIMATES.				7%	NO ESTIMATES	
HCFC-22						15.0	1,500
HCFC-123						1.6	87
HCFC-134a						16.0	1,300
HCFC-143a						41.0	2,900
HCFC-152a						1.7	140

* = parts per million
** = parts per billion

[1] V. Ramanathan et al., "Trace Gas Effects on Climate," in *Atmospheric Ozone 1985*, Global Ozone Research and Monitoring Project Report No. 16 (Washington, DC: National Aeronautics and Space Administration, World Meteorological Organization, 1985).

[2] Intergovernmental Panel on Climate Change, *Scientific Assessment of Climate Change*, Summary and Report, World Meteorological Organization/UN Environment Program (Cambridge, MA: Cambridge University Press, 1990).

[3] U.S. Environmental Protection Agency, Office of Policy Planning and Evaluation, *Policy Options for Stabilizing Global Climate*, Draft Report to Congress (Washington, DC: Government Printing Office, June 1990).

Source: Adapted from U.S. Congress, Office of Technology Assessment, *Changing by Degrees: Steps To Reduce Greenhouse Gases*, OTA-O-482 (Washington, DC: February 1991), pp. 54–55.

Table 4.5.
Estimated Carbon Emissions as a Result
of Deforestation (in millions of tons)

Nation	Estimated Emissions	Nation	Estimated Emissions
Brazil	336	Laos	85
Indonesia	192	Nigeria	60
Colombia	123	Philippines	57
Ivory Coast	101	Burma	51
Thailand	95	Malaysia	50
Others	514		

Source: R. A. Houghton et al., "The Flux of Carbon from Terrestrial Ecosystems to the Atmosphere in 1980 Due to Changes in Land Uses: Geographic Distribution of Global Flux," *Tellus*, February–April 1987 as reprinted in Christopher Flavin, *Slowing Global Warming: A Worldwide Strategy* (Washington, DC: Worldwatch Institute, 1989), p. 29.

This is probably two orders of magnitude greater than the usual rate of carbon dioxide production from volcanoes, which on the average must be equal to the rate at which silicates are weathered to carbonates. Thus humans are now carrying out a large scale geophysical experiment of a kind that could not have happened in the past nor be reproduced in the future. Within a few centuries we are returning to the atmosphere and oceans the concentrated organic carbon stored in sedimentary rocks over hundreds of millions of years. This experiment, if adequately documented, may yield a far-reaching insight into the processes determining weather and climate.

Predictions of Global Warming in the Late 1980s

During the 1980s, some scientists began to speak more forcefully about the possibilities of anthropogenically induced global warming. The unusually hot summer of 1988 appeared to draw the attention of politicians and the general public to this topic. During hearings of House and Senate committees, a few scientists forcefully pointed out the possible threats of climate change.

This is the testimony of Dr. James Hansen of NASA's Goddard Institute of Space Studies before the Senate Committee on Energy and Natural Resources, June 23, 1988:

I would like to draw three main conclusions. Number one, the earth is warmer in 1988 than at any time in the history of instrumental measurements. Number two, the global warming is now large enough that we can ascribe with a high degree of confidence a cause and effect relationship to the greenhouse effect. And number three, our computer climate simulations indicate that the greenhouse effect is already large enough to begin to effect the probability of extreme events such as summer heat waves. . . .

Altogether the evidence that the earth is warming by an amount which is too large to be a chance fluctuation and the similarity of the warming to that expected from the greenhouse effect represents a very strong case. [It is] my opinion that the greenhouse effect has been detected, and it is changing our climate now.

Here is the testimony of Dr. Wallace S. Broecker of Lamont-Doherty Geological Observatory of Columbia University before the Subcommittee on Environmental Protection and Hazardous Wastes and Toxic Substances of the Senate Committee on Environment and Public Works, January 28, 1987:

The inhabitants of planet Earth are quietly conducting a gigantic environmental experiment. So vast and so sweeping will be its impacts that, were it brought before any responsible council for approval, it would be firmly rejected as having potentially dangerous consequences. Yet the experiment goes on with no significant interference from any jurisdiction or nation. The experiment in question is the release of CO_2 and other so-called greenhouse gases to the atmosphere. As these releases are largely by-products of energy and food production, we have little choice but to let the experiment continue. We can perhaps slow its pace by eliminating frivolous production and by making more efficient our use of fossil fuel energy. However, beyond this we can only prepare ourselves to cope with the impacts the greenhouse buildup will bring. . . .

My impressions are more than educated hunches. They come from viewing the results of experiments nature has conducted on her own. The results of the most recent of these experiments are well portrayed in polar ice, in ocean sediment, and in bog mucks. What these records tell me is that Earth's climate does not respond in a smooth and gradual way; rather it responds in sharp jumps. These jumps appear to involve large-scale reorganizations of the Earth's system. If this reading of the natural record is correct, then we must consider the possibility that the major responses of the system to our greenhouse provocation will come in jumps whose timing and magnitude are unpredictable. Coping with this type of change is clearly a far more serious matter than coping with a gradual warming.

In response to Dr. Broecker's testimony, Senator Mitchell asked what time frame might be expected for the next climatic "jumps." Dr. Broecker responded that:

if the climate system does change mainly through reorganization, then it would be fair to say that a jump should occur sometime during the next 100 years. Our preparation for such an eventuality must be a continuing effort. As with cancer, we are not likely to gain the answers we seek in the next 50 years.

Differences of Opinion

Whether they agreed that global warming had actually begun, most scientists in the late 1980s appear to have agreed that the conditions needed to bring about that warming were already in place. A number of authorities were disturbed, however, by the case that was being presented for global warming. They pointed out that the effects of global climatic change were largely unknown, and, therefore, they urged caution, called for more research, and complained about the "media hype" surrounding public debate over global warming. The following is a letter to President George Bush from Dr. Patrick J. Michaels, University of Virginia, and Dr. Robert C. Balling, Jr., Arizona State University, February 1, 1991:

> In October 1990, thirty environmental scientists of national and international reputation met in Phoenix for a symposium sponsored by the Laboratory of Climatology at Arizona State University to address the problem of global climatic change. Together they produced a Research Agenda titled "Global Climatic Change: A New Vision for the 1990s."
>
> While all of these scientists agree that the Earth will undergo some warming, they are concerned about the mounting and compelling evidence that the "popular vision" of apocalyptic climate change—a global temperature rise of 4° C, disastrous sea level rise, and civil strife resulting from ecological chaos—is dramatically distorted. Our concern becomes more urgent as calls for economically significant remedies become more strident.
>
> The working hypothesis of this group of scientists is that the impact of future climatic change is likely to be much less detrimental than generally expected, perhaps even neutral or beneficial.

The letter then cites lines of evidence to support this position:

> It is truly ironic that the highly politicized nature of the global warming problem tends to suppress support for research that may lead to a more balanced view of the future. Most "global change" programs have bloomed because of the specter of an apocalypse. They cannot rationally be expected to be enthusiastic about research that compromises their "raison d'être."
>
> We urge you to bring the enclosed research documents to the attention of appropriate federal research agencies as well as the private sector, which may be more disposed to support research that could lead to a balanced assessment of the issue. This is especially

true when a premature attempt to deal with, what in fact may be, a misinterpreted problem could lead to precipitous economic dislocations.

Here is the testimony of Dr. Andrew Solow, statistician at Woods Hole Institute of Oceanography, before the Subcommittee on Oceanography and the Great Lakes of the House Committee on Merchant Marine and Fisheries, May 4, 1989:

> Unfortunately, substantial uncertainties exist about climate change. Much of our information about climate change comes from experiments with climate models. The results of these experiments are sometimes treated inappropriately, in my opinion, as forecasts of future climates.
>
> Climate models do not perform well at reproducing the recent behavior of climate. For example, these models predict that a global warming of at least 1° C should have occurred over the past 100 years in response to increasing levels of atmospheric carbon dioxide.
>
> Although there appears to have been some warming over that period, it has been no more than half a degree Centigrade, and possibly much less.
>
> Moreover, there is no clear evidence that any warming that has occurred has been due to an enhanced greenhouse effect. In that case, not only have the models erred in predicting past greenhouse warming that did not occur, but they have also erred in not predicting past warming that has occurred for other reasons. . . .
>
> With regard to the question of whether accelerated greenhouse warming due to human activities has begun, it is time to stop waffling. Existing data do not support the conclusion that it has. . . .
>
> On the basis of current information, I believe that the benefits of further research are very great, and that postponing a more active policy response, even by several years, is very unlikely to alter the ultimate outcome of climate change.

The Consensus View

It's not clear that any single statement can represent the divergent views about global warming held by scientists. Perhaps the closest to such a statement is the following executive summary (taken from J. T. Houghton, G. J. Jenkins, and J. J. Ephraums, *Climate Change: The IPCC Scientific Assessment,* Executive Summary [Cambridge, MA: Cambridge University Press, 1990]), prepared for a 1990 book on global warming by the Intergovernmental Panel on Climate Change. In this statement, the number of asterisks indicate the relative amount of certainty about each event. The possible range is from five asterisks, indicating virtual certainty, to one asterisk, indicating a low degree of confidence.

Temperature:
***** the lower atmosphere and Earth's surface warm;
***** the stratosphere cools;
  *** near the Earth's surface, the global average warming lies between +1.5° C and +4.5° C, with a "best guess" of 2.5° C;
  *** the surface warming at high latitudes is greater than the global average in winter but smaller than in summer . . .
  *** the surface warming and its seasonal variation are least in the tropics.

Precipitation:
**** the global average increases (as does that of evaporation), the larger the warming, the larger the increase;
  *** increases at high latitudes throughout the year;
  *** increases globally by 3% to 15% (as does evaporation);
   ** increases at mid-latitudes in winter;
   ** the zonal mean value increases in the tropics although there are areas of decrease . . .
   ** changes little in subtropical arid areas.

Soil moisture:
  *** increases in high latitudes in winter;
   ** decreases over northern mid-latitude continents in summer.

Snow and sea-ice:
**** the area of sea-ice and seasonal snow-cover diminishes.

Politicians Speak

The potential threat of climatic change as a result of human activities was hardly news to scientists in the 1980s. Many had been thinking and working on the problem for decades. But the unusually hot decade of the 1980s suddenly made global warming an issue of concern to nonscientists as well. As was the case among scientists, however, spokespersons differed as to the potential risk assigned to the increasing release of greenhouse gases. An example is this statement of Senator J. Bennett Johnston of Louisiana before the Senate Committee on Energy and Natural Resources on June 23, 1988:

> Last November, we had introductory hearings on the question of global warming and the greenhouse effect. We listened with mixtures of disbelief and concern as Dr. Manabe told us that the expected result of the greenhouse effect was going to be a drying of the southeast and midwest. Today as we experience 101° temperatures in Washington, DC, and the soil moisture across the midwest is ruining the soybean crops, the corn crops, the cotton crops, when we're have emergency meetings of the Members of the Congress in order to figure out how to deal with this emergency, then the words of Dr. Manabe and other witnesses who told us about the greenhouse effect are becoming not just concern, but alarm. . . .

So, as we begin today, we are doing so with a consciousness that this is not some esoteric study of little interest to the ordinary citizen of the United States. This is not some economic study on somebody's theory. The greenhouse effect has ripened beyond theory now. We know it is fact. What we don't know is how quickly it will come upon us as an emergency fact, how quickly it will ripen from just simply a matter of deep concern to a matter of severe emergency.

Another view of the global warming issue was presented by then White House Chief of Staff John Sununu in a statement before the National Academy of Engineering in the fall of 1989:

Although I agree that [global warming] is a critical issue, the fact is that the models with which analysis is being done and with which policy is being moved, as good as they may be, still are based on element sizes measured in hundreds of kilometers in length and width, and tens of kilometers in thickness. I suspect that no one who has ever been involved in engineering simulation would feel comfortable making major decisions in which the elements were orders of magnitude greater than the details on which they were looking for information. And yet the fact is that we are moving toward binding international policy based on conclusions being drawn by policymakers who have no sense at all of the difference between the levels of confidence they want to have. A system is not valid just because it gives you the answers you want. And yet so much policy is being made in reaction to that principle.

Future Scenarios

The two fundamental predictions scientists would like to make concern (1) changes in the concentration of carbon dioxide and other greenhouse gases in the atmosphere and (2) global temperature changes that may result from these. A number of authorities have made one or both types of calculations. Two examples of these are shown in Figures 4.11 and 4.12.

Figure 4.11 shows the amount of carbon dioxide that would be retained in the atmosphere under four different scenarios. In the "very low scenario," the world would actually decrease the amount of carbon dioxide released to the atmosphere every year. In the "low scenario," the annual growth in carbon dioxide emissions would range between 1.0 percent and 1.49 percent. The annual growth in emissions would be 1.5 percent to 1.79 percent for the "medium scenario," and greater than 1.79 percent for the "high scenario."

Figure 4.12 shows estimated changes in the annual global temperature assuming the most likely changes in carbon dioxide

Figure 4.11.

Final Retention of Carbon Dioxide in the Atmosphere under Four Scenarios (retention is measured in additional tons of CO_2 retained in the atmosphere compared with the tons of CO_2 emitted)

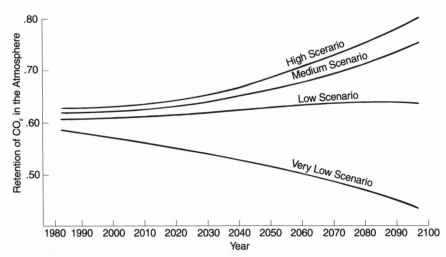

Source: Stephen Seidel and Dale Keyes, *Can We Delay a Greenhouse Warming?* 2d ed. (Washington, DC: Environmental Protection Agency, Office of Policy, Planning, and Evaluation, Office of Policy Analysis, 1983), pp. 3–21.

Figure 4.12.
Estimated Global Mean Temperature Changes under Three Conditions of Climate Sensitivity

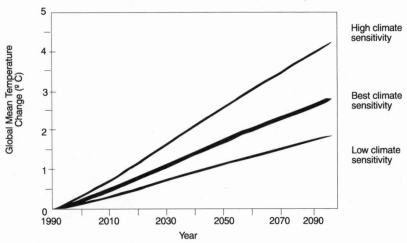

Source: Intergovernmental Panel on Climate Change, *1992 IPCC Supplement,* photocopied report, February 1992, p. 25.

concentrations in the atmosphere, as determined by scientists working for the Intergovernmental Panel on Climate Change in 1992. The three lines shown in this graph represent uncertainty as to how the global temperature will respond to increases in carbon dioxide concentrations. Notice that the estimated temperature increase by 2090 varies from a low of 2° C (3.6° F) to a high of 4° C (7.2° F).

Possible Impacts of Global Warming

Global warming has become an important international issue because of the impact a warmer climate could have on many aspects of the environment and on human societies. Various authorities have suggested that a higher annual average temperature could produce rising sea levels, improved agricultural conditions in some areas and poorer conditions in others, a loss of water resources in some places and augmented resources in others, damage to human health, damage to or destruction of urban infrastructure, and so on. The selections below summarize a few of these predictions.

Recall that all of the predictions are based on existing conditions, future trends, and natural processes that often are not well understood. For each prediction, there are likely to be critics who can find fault with the scenario described.

Sea-Level Changes

One of the most common predictions is that sea levels will rise. One reason for this change is the melting of ice caps and glaciers that would result from higher global temperatures. A more important factor, however, is the expansion of seawater that results from an increase in temperature. Rising sea levels mean, among other things, that shorelines throughout the world will change. For example, some Pacific islands are no more than a meter above sea level today. Higher sea levels would submerge some island nations entirely.

The changes that might result from an increase in sea levels have been described by James G. Titus, senior scientist with the U.S. Environmental Protection Agency (see James G. Titus, ed., *Effects of Changes in Stratospheric Ozone and Global Climate Change* [Washington, DC: Environmental Protection Agency, 1986], pp. 286–287). Some of those changes include the following:

- A substantial rise in sea level would permanently inundate wetlands and lowlands, accelerate erosion, exacerbate coastal flooding, and increase the salinity of estuaries and aquifers.

- Bangladesh and Egypt appear to be among the nations most vulnerable to the rise in sea level projected for the next century. Up to 20% of the land in Bangladesh could be flooded with a 2-meter rise in sea level. Although less than 1% of Egypt's land would be threatened, over 20% of the Nile Delta, which contains most of the nation's people, would be threatened.

- Erosion caused by sea level rise could threaten recreational beaches throughout the world. Case studies have concluded that a 30-centimeter rise in sea level would result in beaches eroding 20 to 60 meters or more. . . .

- Sea level rise would increase the costs of flooding, flood protection and flood insurance in coastal areas. . . .

- Increased salinity from sea level rise would convert cypress swamps to open water and threaten drinking water supplies. . . .

- River deltas throughout the world would be vulnerable to a rise in sea level, particularly those whose rivers are dammed or leveed.

Water Resources

Climate models predict that the amount of moisture in the soil will change by different amounts and in different ways in various parts of the world. Figure 4.13 shows the predictions produced by the general circulation model at NOAA's Geophysical Fluid Dynamics Laboratory at Princeton, New Jersey. In Figure 4.13, the shaded areas represent areas predicted to have a decrease in soil moisture; the clear areas represent areas predicted to have an increase in soil moisture. Numbers are percent decreases or increases of soil mosture. As Figure 4.13 shows, soil moisture would increase in tropical regions (up to 100 percent in India, for example) and decrease in all parts of North America, Europe, and northern Africa.

Agriculture

Increased levels of carbon dioxide in the atmosphere may have both beneficial and harmful effects on plant life on Earth. The

Figure 4.13.
**Projected Changes in Soil Moisture in Response to a Doubling of
CO₂ in the Atmosphere (Shaded areas represent areas predicted to
have a decrease in soil moisture; clear areas represent areas predicted
to have an increase in soil moisture. Numbers are percent decreases
or increases of soil moisture.)**

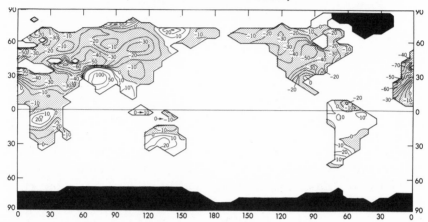

Source: S. Manabe and R. T. Wetherald, "Reduction in Summer Soil Wetness Induced by an Increase in Atmospheric Carbon Dioxide," *Science,* May 2 1986, p. 627.

following is a summary of some of these predicted effects (taken from Susan C. Whitmore, *Global Climate Change and Agriculture: A Summary* [Beltsville, MD: Department of Agriculture, 1991], p. 4):

Although climate change is not likely to threaten U.S. food supplies, it may:

- Reduce or increase average yields of corn, soybeans, and wheat, both rain fed and irrigated. Increased yields may occur in northern latitudes where warmer conditions would provide a longer frost-free growing season. Decreased yields may occur in mid-latitudes primarily from higher temperatures which shorten a crop's life cycle.

- Increase the yields of corn, soybean, and wheat since increased atmospheric CO₂ may increase plant growth.

- Result in a northward shift in cultivated land, causing significant regional dislocations in agriculture with associated impacts on regional economies.

- Shift rainfall patterns, which might expand crop irrigation requirements in certain regions and, hence, increase competition for regional water supplies and increase surface and ground water pollution.

- Change the ranges and populations of agricultural pests. These effects could change pest control requirements.

- Result in sea level rise and the flooding of near-coastal agricultural lands.

Forests

The interaction between forests and atmospheric carbon dioxide is not entirely understood. There has been a great deal of controversy as to the effects on forests worldwide as a result of increasing levels of carbon dioxide in the atmosphere. The summary below represents one scenario as to what might happen to the forests that cover about one-third of U.S. land area (taken from Susan C. Whitmore, *Global Climate Change and Agriculture: A Summary* [Beltsville, MD: Department of Agriculture, 1991], p. 4):

- The southern ranges of many forest species in the eastern United States could experience dieback of several hundred miles depending on the extent of temperature changes and drying soils. This dieback could result in a serious loss of productivity, depending on how dry the soils become.

- The potential northern range of forest species in the eastern United States could shift northward as much as 400 miles over the next century. Productivity could increase along the northern limits of some eastern species.

- If elevated CO_2 concentrations substantially increase tree growth through an increase in photosynthesis and efficiency of water use, the southern declines could be alleviated.

- If climate stabilizes, forests might eventually regain a generally healthy status over a period of several centuries. In the meantime, declining forests could be subject to increased fires, pest attacks, disease outbreaks, wind damage, and air pollution.

- Additional possible impacts of changes in forests include reductions in biotic diversity, increased soil runoff and erosion, reduced aquifer recharge, and changes in wildlife habitat and recreation.

Ecological Systems

Biologists know a great deal about the effects of climate on various plants and animals. The following passage summarizes some potential effects of global warming on animal species (taken from Robert L. Peters, "Effects of Global Warming on Biological Diversity," from the First North American Conference on Preparing for Climate Change: A Conservative Approach, Washington, DC, October 27–29, 1987, as quoted in Dean Edwin Abrahamson, ed., *The Challenge of Global Warming* [Washington, DC: Island Press, 1989], pp. 82ff):

> If the planet warms as projected, natural ecosystems would be stressed by large changes in temperature, moisture patterns, evaporation rates, and other associated chemical and physical changes.
>
> We can infer how the biota might respond by observing the world as it is today. By observing present distributions of plants and animals, which are determined in large part by temperature and moisture patterns, it is possible to hypothesize what would happen if these underlying temperature and moisture patterns changed.
>
> For example, if we know that one race of the dwarf birch, *Betula nana*, can only grow where the temperature never exceeds 22° C [citation given], then we can hypothesize that it would disappear from those areas where global warming causes temperatures to exceed 22° C. . . .
>
> These kinds of observations tell us that plants and animals are very sensitive to climate. Their ranges move when the climate patterns change—species die out in areas where they were once found and colonize new areas where the climate becomes newly suitable. We also know from the fossil record that some species have become completely extinct because they were unable to find suitable habitat when climate change made their old home unlivable. . . .
>
> Finally, not all the changes for wild systems would be negative. Some species would expand their ranges and have greater abundances. . . .
>
> However, the most optimistic thing that should be said about the future of natural systems under a regime of warming climate is that a great deal of rearrangement would occur, and it is most likely the outcome will be widespread extinction of species.

Human Societies

Climate change very well may have an impact on human societies, particularly on human health and on the physical structures that are part of our urban areas. The selections below outline some of these possible impacts (taken from Joel B. Smith and Dennis A.

Tirpak, eds., *The Potential Effects of Global Climate Change on the United States* [New York: Hemisphere, 1990], pp. 525–526, 557):

Human Health
• Although there may be increases in weather-related summer deaths due to respiratory, cardiovascular, and cerebrovascular disease, there may be decreases in weather-related winter deaths for the same diseases. In the United States, however, on average, weather-related deaths are greater in summer than in winter. Thus, global warming could produce a net increase in deaths.
• Sudden changes in temperature are correlated with increases in deaths. So if climatic variability increases, morbidity and mortality are also likely to increase. A decrease in the frequency or intensity of climate extremes is likely to be associated with a decrease in mortality and morbidity.
• Seasonal variation of fetal and infant mortality (higher in the summers, lower in the winters) is thought to be due to summertime increases in infections. The longer and hotter summers that may accompany climate change may exacerbate this effect at least in some regions.
• Vector-borne diseases, such as those carried by ticks, fleas, and mosquitos, could increase in certain regions. In addition, climate change may alter habitats. For example, some forests may become grasslands, thereby modifying the incidence of vector-borne diseases.
• While uncertainties remain about the magnitude of other effects, it's likely that climate change could have the following impacts:

—Increases in summer rainfall would multiply the amount of ragweed growing on cultivated land.

—Changes in humidity may heighten the incidence and severity of skin infections and infestations such as ringworm, candidiasis, and scabies.

—Increases in the persistence and level of air pollution episodes associated with climate change will have adverse health effects.

Urban Infrastructure
Global climate change could require U.S. cities to make major changes in capital investments and operating budgets. Areas most likely to be affected include water supplies, roads, and bridges, storm sewers and flood control levees, and energy demand in municipal buildings and schools. . . .
Northern cities, such as Cleveland, may incur a change in the mix of their expenditures. In such locations, increased electricity costs for air conditioning could be offset by reductions in expenditures for heating fuel, snow and ice control, and road maintenance. Southern cities could see increases in operating budgets due to the demand for additional air conditioning. . . .
Coastal cities may be subject to more severe impacts. These include 12 of the 20 largest metropolitan areas. For example:

—Sea level rise or more frequent droughts would increase the salinity of shallow coastal aquifers and tidal surface waters. Cities that rely on water from these sources need to examine water supply options. Such areas as Dade County, Florida, or New York City will probably be vulnerable.

—As sea level rises, some coastal cities would require levees to hold back the sea or fill to raise the land surface area. In the case of Miami, the cost of these activities might be more than $500 million over the next 50 to 75 years.

Recommendations for Action

Given what we know about global warming, what actions should individuals and nations take about this problem? Both scientists and nonscientists have had a number of suggestions. Those suggestions range all the way from recommendations for doing essentially nothing (except getting used to a warmer climate) to making radical changes in the way humans live on the planet.

Scientists Speak

Given the uncertainty that surrounds the data on global warming, it is hardly surprising that scientists disagree among themselves as to what, if anything, we should do about climate change. The two selections below illustrate some of the more extreme views on this topic. The first selection is a petition circulated by the Union of Concerned Scientists:

> Global warming has emerged as the most serious environmental threat of the 21st century. There is broad agreement within the scientific community that amplification of the earth's natural greenhouse effect by the buildup of various gases introduced by human activity has the potential to produce dramatic changes in climate. The severity and rate of climate change cannot yet be confidently predicted, but the impacts of changes in surface temperature, sea level, precipitation, and other components of climate could be substantial and irreversible on a time scale of centuries. Such changes could result in severe disruption of natural and economic systems throughout the world.
>
> More research on global warming is necessary to provide a steadily improving data base and better predictive capabilities. But uncertainty is no excuse for complacency. In view of the potential consequences, actions to curb the introduction of greenhouse gases, including carbon dioxide, chlorofluorocarbons, methane, nitrogen

oxides, and tropospheric ozone, must be initiated immediately. Only by taking action now can we insure that future generations will not be put at risk.

The United States bears a special responsibility to provide leadership in the prevention of global warming. It is the world's largest producer of greenhouse gases, and it has the resources to make a great contribution. A thoughtful and vigorous U.S. policy can have a direct, beneficial effect and set an important example for other nations.

The United States should develop and implement a new National Energy Policy, based on the need to substantially reduce the emission of carbon dioxide, while sustaining economic growth. The cornerstones of this policy should be energy efficiency and the expansion of clean energy sources.

The policy should include:

1. A steady increase in motor vehicle fuel economy standards, while the search continues for fuels and other technologies that mitigate carbon dioxide impact;

2. A substantial increase in federal funding for research on energy efficient technologies, as well as federal activities to enhance the adoption of more efficient energy use;

3. Development, demonstration, and commercialization of renewable energy technologies on a massive scale;

4. A nuclear energy program that emphasizes protection of public health and safety, resolution of the problem of radioactive waste disposal, and stringent safeguards against the proliferation of nuclear material and technology that can be applied to weapons construction; and

5. Full consideration of environmental, social, and economic impacts in the establishment of federal subsidies and regulatory standards for development of energy sources.

These measures, along with others designed to curtail the use of chlorofluorocarbons and promote prudent agricultural and reforestation practices, can form the basis for the lowering of greenhouse gas emissions in the United States and other nations. They will provide other, worthwhile benefits to the nation as well, such as more diverse and flexible energy supplies, reduced dependency on imported oil, and the creation of new energy technologies for export and sale in the international marketplace.

Signed by 52 Nobel Laureates, 67 National Science Medalists, and over 700 members of the National Academy of Sciences. The petition was presented to President George Bush on February 6, 1990.

The George C. Marshall Institute is an organization dedicated to providing scientific and technical advice on matters that have impact on public policy. The institute has produced three major reports on global warming. The following is taken from Chapter 6, "Policy Implications," of the institute's 1992 *Global Warming Update: Recent Scientific Findings* (pp. 27–28):

Recent findings, based on observations of actual temperature changes, suggest that the greenhouse warming will be considerably smaller than commonly accepted estimates based on computer simulations. . . .

How do the new results affect energy policy? Some scientists and policymakers want the U.S. to adopt laws severely restricting carbon dioxide emissions, because they regard carbon dioxide as the primary cause of global warming. Congress has asked the Department of Energy for an estimate of the cost of policies that would reduce carbon dioxide emissions by 20% in the next 10 years. According to the Department of Energy, the cost at the end of the decade can be as much as $95 billion/year. The cost of electricity would double. The cost of oil would increase by $60/barrel, and gasoline would go up $1.30/gallon. A privately funded study estimates an accumulated cost of $3.6 trillion over the next 100 years for comparable restrictions.*

But the scientific evidence does not support a policy of carbon dioxide restrictions with its severely negative impact on the U.S. economy. Important new findings on the greenhouse effect and global warming are reported nearly every month. Several of the major findings discussed in this report were released in the last year. Suppose policymakers wait five years to get still more results, before undertaking the drastic measures proposed by concerned scientists and politicians. What will that cost the U.S.?

The Marshall panel did a study on this problem, using data from the 1990 report of the U.N. Intergovernmental Panel on Climate Change. M. E. Schlesinger and X. Jiang did a similar study.* Both studies yielded the same answer. A five-year delay on major policy decisions regarding carbon dioxide limits will lead to a small amount of warming in the next century. How small will the additional warming be?

The calculations show that a five-year delay in limiting carbon emissions will make the world warmer in the next century by at most one tenth of a degree, compared to how warm it would be if there were no delay. [italics in original]

A very rapid evolutionary process is occurring in the field of greenhouse research, with major improvements likely in basic understanding and in the accuracy of the greenhouse forecasts in the next few years. An additional warming of one tenth of a degree in the 21st century is a very small penalty to pay for better information on government decisions that, if take unwisely, can be extraordinarily costly to the U.S. economy.

* Endnotes at these points have been omitted.

Politicians Speak

Of course, political leaders have taken as wide a variety of positions on global warming as have scientists. The following excerpts illustrate the range of views expressed by those who make policy. The first selection reflects the views of Senator George Mitchell, majority leader of the U.S. Senate, as to what the world's nations must do to deal with climate change (taken from George J. Mitchell, *World on Fire: Saving an Endangered Earth* [New York: Charles Scribner's Sons, 1991], pp. 215–218).

> • They must, no matter how difficult it is or how long it takes, negotiate, sign, ratify, and implement an international accord for global protection. . . .
> • They must find a satisfactory way to pay for the environmental war the nations must now wage together. . . .
> • They must immediately convene a greenhouse summit of the world's nations to develop and implement a strategy. . . .
> • They must launch a major initiative, spearheaded by a prestigious world panel, to develop a long-term strategy to fulfill national and world economic needs without destroying the global environment. . . .
> • They must refine research and find better ways to more accurately monitor the planet's vital signs. . . .
> • They must launch a sustained international program to develop alternative fuels that do not generate carbon dioxide and the other greenhouse gases.
> • They must reforest the earth. . . . This effort is not moving fast enough.

Other politicians are less eager to move forward on climate issues, preferring to resolve some of the uncertainty regarding global warming. The position of the Reagan and Bush administrations from 1980 to 1992 was consistently that governments should not act on global warming until a good deal more research had been done. That position is outlined in the following speech by President George Bush before the Intergovernmental Panel on Climate Change on February 5, 1990:

> We all know that human activities are changing the atmosphere in unexpected and unprecedented ways. Much remains to be done. Many questions remain to be answered. Together, we have a responsibility to ourselves and the generations to come to fulfill our stewardship obligations. But that responsibility demands that we do it right. We acknowledge a broad spectrum of views on these issues, but our respect for a diversity of perspective does not diminish our recognition of our obligation or soften our will to produce policies

that work. Some may be tempted to exploit legitimate concerns for political positioning. Our responsibility is to maintain the quality of our approach, our commitment to sound science, and an open mind to policy options.

So the United States will continue its efforts to improve our understanding of climate change—to seek hard data, accurate models, and new ways to improve the science—and determine how best to meet these tremendous challenges. Where politics and opinion have outpaced science, we are accelerating our support of the technology to bridge that gap. And we are committed to coming together periodically for international assessments of where we stand. . . .

Our goal continues to be matching policy commitments to emerging scientific knowledge and a reconciling of environmental protection to the continued benefits of economic development. And as Secretary [of State] Baker observed a year ago, whatever global solutions to climate change are considered, they should be as specific and as cost-effective as they can possibly be. If we hope to promote environmental protection and economic growth around the world, it will be important not to work in conflict but with our industrial sectors. That will mean moving beyond the practice of command, control, and compliance towards a new kind of environmental cooperation and toward an emphasis on pollution prevention rather than mere mitigation and litigation.

Adapting to Change

In the range of policy options, the simplest approach is to "grin and bear it." Some authorities argue that climate change has always taken place. The easiest, most economical solution may be for humans simply to learn to live with a slightly warmer world. The following passage explains the idea of adaptation to climate change (taken from Pierre R. Crosson and Norman J. Rosenberg, "Adapting to Climate Change," in *Carbon Dioxide Emissions and Global Warming* [Washington, DC: Resources for the Future, Spring 1991], p. 19):

It is also important to recognize that there are two kinds of adaptive response to climate change, to which we now turn. One response includes all those things people would be induced to do within the existing institutional and policy regime. The other consists of institutional and policy changes that would be called for where and when the existing regime proved inadequate to deal with the impacts of climate change. . . .

Examples of the two kinds of adaptive response to climate change can be found in agriculture. Studies of the impacts of climate change on agriculture show that in many areas, including the U.S. Midwest, crop yields (output per acre) might fall with higher temperatures and less precipitation. The fall in yields would increase production costs to farmers, inducing them to investigate existing technologies and management practices for better ways to adapt to the changed climate. Farmers might turn to conservation tillage, a technique that conserves more soil moisture than the more commonly used tillage techniques. They might also adopt already available crop varieties that are better adapted to the hotter and drier climate, and invest in irrigation to counter the decline in precipitation. All of these adaptations are examples of measures that people would be induced to undertake within the existing institutional and policy regime.

However, in some circumstances, these induced adaptations may be judged inadequate in the sense that after they have been made, society appraises the remaining costs of climate change as unacceptably high. In such a case, institutional or policy change would be called for to develop additional adaptations that would bring the remaining costs within acceptable limits.

If farmers find that the alternatives available to them from among existing technologies and management practices are inadequate to compensate for the negative impacts of climate change, they may face the prospect of going out of farming, and perhaps leaving a region altogether. This prospect would stimulate agricultural research institutions and those charged with responsibility for agricultural policy to invest more in research to develop a new set of technologies and practices better adapted to the changed climatic regime. Institutional rules for allocating irrigation water might also be changed to give farmers greater flexibility in using water on their own farms and in transferring it among farmers.

Technological Fixes

One approach to solving the global warming problem is to look for one or more "technological fixes." A technological fix is a scientific method that will decrease the amount of greenhouse gases released to the atmosphere or that will counteract the gases once they reach the atmosphere. Two technological fixes that have been recommended are discussed below.

Carbon Dioxide Scrubbing

This method is described in a 1991 report by the Office of Technology Assessment (*Changing by Degrees: Steps To Reduce Greenhouse Gases,* OTA-O-482 [Washington, DC: Government Printing Office, February 1991], p. 95).

It is also possible to remove CO_2 from flue gases and liquify it—through a process known as "scrubbing." Theoretically, one could pump the liquefied CO_2 through pipelines to disposal sites, for example, the deep ocean, where it is hoped it will remain rather than entering the atmosphere. . . .

Carbon dioxide scrubbing basically involves:

- compressing and cooling the stack gases;

- removing CO_2 from the bases via a reaction with a solvent solution;

- heating and steam-stripping the CO_2-enriched solution to reverse the reaction, yielding uncondensed steam and CO_2;

- condensing and removing water vapor, leaving the recovered CO_2; and

- compressing and liquefying the recovered CO_2.

Unfortunately, this technology is too expensive to use at this point. The following statement explains why (taken from Christopher Flavin, *Slowing Global Warming: A Worldwide Strategy*, Worldwatch Paper 91 [Washington, DC: Worldwatch Institute, 1989], p. 33):

> The first of these options [carbon dioxide scrubbing] can be discarded relatively quickly, despite its superficial appeal. While the removal of up to 90 percent of the carbon could in theory be accomplished at large power plants, . . . these are unproven technologies. Unlike the sulfur that causes acid rain, carbon is not a minor contaminant of fossil fuels; rather it makes up 73 percent of the weight of the average coal. Huge quantities of it would have to be extracted and disposed of. The average 500-megawatt power plant releases 100 million kilograms of carbon each year, or seven times the amount of ash currently trapped in precipitators. The only way that this much carbon could be kept from the atmosphere is to bury it in the deep oceans, something that would not only be costly but would require the construction of long pipelines to carry it to the sea.

Enriching the Oceans with Iron

Another suggestion has been to add iron-rich dust to the oceans. The selections below describe the theory and method.

A great deal of uncertainty surrounds this process, however, because scientists have no idea what side effects might occur. As the inventor of the process has said, "It's scary. I don't want to go down in history as Martin's Mistake—as this guy who advocated adding iron, and they did it, and it completely ruined the ecosystem." The first selection below (taken from John H. Martin, R. Michael Gordon, and Steve E. Fitzwater, "Iron in Antarctic Waters," in *Nature*, May 10, 1990, p. 158) outlines the scientific concept behind this method, whereas the second selection (taken from Robert Kunzig, "Earth on Ice," *Discover*, April 1991, p. 59) describes how that concept might be applied in practice:

In conclusion, the results presented here clearly support the observations of the 1925–27 *Discovery* Expedition scientists—that the neritic areas probably were (and are) rich in phytoplankton because of Fe [iron] introduced from land.* The very low dissolved Fe (and Mn [manganese]) levels in the Drake passage provide support for our argument* that present-day plant productivity is limited in offshore waters of the Antarctic because of Fe deficiency. Judging by the unused excess of major plant nutrients, this lack of essential Fe* seems to be severely limiting the power of the "biological pump" and thus contributing to the raised atmospheric CO_2 concentrations typical of previous and present interglacial periods (preindustrial ≈ 280 p.p.m.).* In contrast, greatly enhanced Fe input from atmospheric dust* may have stimulated phytoplankton growth and increased the power and efficiency of the biological pump, thus contributing to the drawing down of atmospheric CO_2 during glacial maxima.*

* Endnotes omitted.

If adding iron to a bottle causes the phytoplankton in the bottle to bloom, then you ought to be able to fertilize the ocean itself. It wouldn't take very much iron to fertilize the whole Southern Ocean, because phytoplankton don't need very much. Martin did a back-of-the-envelope calculation. If 300,000 tons of iron were broadcast in the ocean surrounding Antarctica over a six-month growing season, he figured, it would allow the phytoplankton to convert all the available nutrients—including 2 billion tons of carbon—into new organic matter. Those 2 billion tons would come from carbon dioxide that had dissolved out of the atmosphere into the surface waters. In other words, they would come right out of our greenhouse. . . .

"You give me half a tanker full of iron," Martin said this past May, . . . "and I'll give you another ice age."

Social and Political Action

Most people feel that political action, not some technological fix, is necessary to reduce the amount of greenhouse gases in the atmosphere. For example, individuals and nations could be encouraged to improve the efficiency of their carbon-burning devices. Or they could be encouraged to reduce the cutting of trees, to begin planting more trees, or both. Some of these suggestions are reflected in the following sections.

Energy Conservation

A straightforward method for reducing emissions of carbon dioxide is to cut back on the amount of fossil fuels we burn. This method does not depend on the development of new technologies and can be implemented whenever people are motivated to do so. Table 4.6 shows the savings and costs of various energy conservation programs for this nation as estimated by the Office of Technology Assessment of the U.S. Congress. These predictions were based on a "modest" and a "tough" scenario, as defined below (taken from U.S. Congress, Office of Technology Assessment, *Changing by Degrees: Steps To Reduce Greenhouse Gases*, OTA-O-482 [Washington, DC: Government Printing Office, February 1991], pp. 5, 318–319, 321):

> Energy conservation is the logical first step for the United States if it wishes to reduce its own CO_2 emissions below present levels over the next 25 years. For comparison, if no actions are taken, emissions of CO_2 will likely rise 50 percent during the next quarter century. Under a set of modest policies, designed to encourage people to choose technologies that are cost-effective, emissions of CO_2 probably will rise about 15 percent over the next 25 years. This policy package is labeled OTA's "Moderate" scenario.

OTA also identified an energy conservation, energy-supply, and forest-management package that can achieve a 20 percent to 35 percent emissions reduction. This package is labeled OTA's "tough" scenario. Examples of changes that would be included in a "tough" scenario include efficiencies for automobiles of 55 miles per gallon by 2010 and building new homes and retrofitting old homes to reduce heat use by 85 percent.

OTA then tries to provide a dollar amount on the costs of implementing its "tough" scenario. Its conclusions are as follows:

> *Total Costs (by 2015)*
> Adding results documented below by sector yields the following net annual costs (i.e., annualized capital and operating costs minus fuel savings):
> Utilities: +$35 billion
> Residential buildings: −$25 to −$15 billion
> Commercial buildings: −$28 to +$22 billion
> Transportation: −$35 to +$38 billion
> Industry: +$21 to +$58 billion
> Forestry: +$10 to +$13 billion
> TOTAL: −$22 to +$150 billion

Energy conservation seems to be nearly everyone's first choice for dealing with climate change. But a few observers, though applauding the idea of conservation, ask whether it is realistic to expect Americans and those in other developed nations to change their lifestyles to accomplish this objective. The following selection

Table 4.6
Measures to Lower U.S. Carbon Emissions

	2000		2015	
	Moderate	Tough	Moderate	Tough
DEMAND-SIZE MEASURES				
Residential Buildings				
New investments:				
Shell efficiency	0.5%	0.7%	1.3%	2.0%
Heating and cooling efficiency	0.0%	0.2%	0.1%	0.4%–0.6%
Water heaters and appliances	0.5%	0.5%	1.2%	1.5%–2.3%
O&M retrofits:				
Shell efficiency	0.6%	1.7%	0.8%	0.9%
Lights	0.4%	0.6%	0.6%	0.8%
All residential measures together	1.9%	3.7%	3.9%	5.6%–6.6%
Commercial Buildings				
New investments:				
Shell efficiency	0.9%	1.4%	2.3%	4.0%
Heating and cooling efficiency	0.4%	0.5%–1.0%	1.0%	1.2%–1.9%
Lights	0.8%	1.1%	2.1%	3.0%
Office equipment	0.5%	0.7%	1.6%	2.1%
Water heaters and appliances	0.1%	0.1%	0.1%	0.1%
Cogeneration	0.1%	0.4%–0.6%	0.2%	1.5%–2.3%
O&M retrofits:				
Shell efficiency	0.6%	2.1%	0.8%	0.8%
Lights	0.2%	0.2%	0.5%	0.5%
All commercial measures together	3.5%	6.6%–7.3%	8.5%	13%–15%
Transportation				
New investments:				
New auto efficiency	0.4%	0.8%–1.2%	0.8%	3.5%–3.8%
New light truck efficiency	0.3%	0.5%–0.8%	0.5%	2.5%–2.7%
New heavy truck efficiency	0.2%	0.7%–0.8%	0.4%	2.4%
Non-highway efficiency	0.3%	0.7%	0.5%	1.2%
O&M retrofits:				
Improved public transit	0.1%	2.1%	0.2%	3.5%
Truck inspection & maintenance	0.3%	0.3%	0.3%	0.4%
Traffic flow improvements/				
55 mph	1.2%	1.2%	1.2%	1.4%
Ridesharing/parking controls	0.3%	0.5%	0.4%	1.0%
All transportation				
measures together	3.1%	7.0%–7.8%	4.2%	14%–15%
Industry				
New investments:				
Efficient motors	0.3%	1.0%–1.1%	1.2%	3.7% -4.0%
Lighting	0.1%	0.2%	0.5%	0.7%–0.8%
Process change, top 4 industries	0.7%	2.2%	3.0%	8.2%
Fuel switch to gas	0.0%	0.5%–0.7%	0.0%	2.4%–2.7%
Cogeneration	0.3%	1.1%–1.4%	0.8%	5.2%–5.8%

	2000		2015	
	Moderate	Tough	Moderate	Tough
O&M retrofits:				
Housekeeping	0.6%	1.5%	1.9%	2.0%
Lighting	0.1%	0.1%	0.1%	0.2%
All industry measures together	2.2%	6.1%–6.7%	7.6%	17%–18%
UTILITY SUPPLY-SIDE MEASURES				
Existing plant measures:				
Improved nuclear utilization	1.3%	1.3%	4.1%	4.1%
Fossil efficiency improvements	1.7%	1.7%	1.7%	1.7%
Upgraded hydroelectric plants	0.6%	0.6%	0.5%	0.5%
Natural gas co-firing	—	3.7%	—	3.7%
New plant measures:				
No new coal; higher fraction of new non-fossil sources	—	0.0%	—	0.0%–4.7%
CO_2 emission rate standards	0.0%	0.0%	0.4%	0.0%–0.1%
All utility measures together	3.4%	6.5%	6.6%	9.9%–14%
FORESTRY MEASURES				
Afforestation:				
Conservation Reserve Program	0.2%	0.2%	0.2%	0.2%
Urban trees	—	0.1%	—	0.7%
Additional tree planting	—	0.6%	—	2.3%
Increased tree productivity	—	0.8%	—	3.1%
Increased use of biomass fuels	—	0.3%	—	1.2%
All forestry measures together	0.2%	2.0%	0.2%	7.5%

Source: U.S. Congress, Office of Technology Assessment, *Changing by Degrees: Steps To Reduce Greenhouse Gases,* OTA-O-482 (Washington, DC: Government Printing Office, February 1991), pp. 318–319.

(taken from Laurence Lippsett, "Wallace Broecker '53: The Grandmaster of Global Thinking," *Columbia College Today,* [no date], p. 25) reflects this view:

> It's an unfortunate thing, but our whole society is geared to high energy use. While most people believe we ought to cut down, it's more a theoretical concern than a reality. Are we really willing to go without air-conditioning in the summer, to drive smaller cars and to drive our cars less often? . . .
>
> I get a little upset because I think we've put an enormous amount of emphasis on conserving energy. . . . But if we're going to have twice as many people on the planet, it seems hard to believe that we're going to conserve to the extent that we're not going to be using more energy in 50 years than we're using now. We're not getting to the root cause of our potential problems for the habitability of the planet, which is population growth. . . .
>
> All of our problems grow with population. This idea that the only way our economy can continue is to have more growth, more jobs,

more products—that's going to blow the lid off. We're accumulating problems in society much faster than we're solving them. The world's leaders must come down hard on population and do everything possible to stabilize it. We must start to think about a steady-state Earth, where we have a stable number of people using a stable amount of energy and a stable amount of products.

Carbon Taxes

Motivating industries and individuals to cut back on carbon emissions can involve some mandatory approaches. One such approach in the case of global warming is called the carbon tax. One form of this option involves taxing a fuel use on the basis of the fuel's net CO_2 emission per unit of energy. The Environmental Protection Agency has conducted a study (taken from Stephen Seidel and Dale Keyes, *Can We Delay a Greenhouse Warming?* 2d ed., Environmental Protection Agency, Office of Policy, Planning, and Evaluation, Office of Policy Analysis [Washington, DC: Government Printing Office, November 1983], p. v) of the potential effect of imposing a carbon tax worldwide and in the United States. Their conclusions were as follows:

- Worldwide taxes of up to 300% of the cost of fossil fuels (applied proportionately based on CO_2 emission from each fuel) would delay a 2° C warming only about 5 years beyond 2040.

- Fossil fuel taxes applied to just certain countries or applied at a 100% rate would not affect the timing of a 2° C rise.

International Action

A number of nations around the world have already begun to develop policies for dealing with the emission of greenhouse gases. Table 4.7 summarizes decisions that had been made as of 1991. As the table shows, many nations have not made final determinations as to the limits they will be placing on carbon dioxide emission or on the target year for reaching these limits.

Alternative Energy Possibilities

A common recommendation for dealing with global warming is the development of alternative forms of energy that do not result in the release of carbon dioxide or other greenhouse gases. In 1988, the Natural Resources Defense Council (NRDC) prepared a recommendation for presidential candidates on this topic. The following is excerpted from that the NRDC statement (taken from

Table 4.7.
Official Greenhouse Gas Emission Stabilization and
Reduction Policies of OECD Countries

Jurisdiction	Base Level Year	Stabilization Year	Percent Reduction	Target Year
Australia	1988	2000	20% of all gases	2005
Austria	1987	ND*	20% of CO_2	2005
Canada	1988	2005	ND	ND
Denmark	1988	2005	20% of CO_2	2000
France	1989/1990	2000	ND	ND
Germany	1987	—	25% of CO_2	2005
Italy	1990	2000	20% of CO_2	2005
Japan	1990	2000	ND	ND
Netherlands	1989/1990	1995	5% of CO_2	2000
New Zealand	1990	ND	20% of CO_2	2005
Norway	1989	2000	ND	ND
Sweden	1988	ND	ND	ND
United Kingdom	1990	2005	ND	ND

*ND = not declared
Source: U.S. Congress, Office of Technology Assessment, *Changing by Degrees: Steps To Reduce Greenhouse Gases*, OTA-O-482 (Washington, DC: Government Printing Office, February 1991), p. 304.

Ralph Cavanagh, David Goldstein, and Robert Watson, *One Last Chance for a National Energy Policy* [Washington, DC: Natural Resources Defense Council, July 1988]):

> The election of Ronald Reagan in 1980 signaled a wholesale retreat from the energy business. The measure of this withdrawal is most cogently captured in dollars: between 1981 and 1987, federal financial support for small hydropower, wind power, solar electricity, and energy from wood and geothermal sources dropped by 80%* . . . .
>
> What was left in the "alternative energy" budget wouldn't buy the booster rockets for a space shuttle, let alone match the effort to land men on the moon. The new administration's official description of its "strategies of national energy policy" began with a pledge "to minimize federal control and involvement in energy markets."* Seldom have presidential aspirations been realized more thoroughly.
>
> *Endnotes omitted

The statement then went on to outline the current status of alternative energy systems:

> *Solar Heating and Cooling Systems:* Installed in less than 1 million of the nation's 80 million households, and rapidly losing momentum. From

a peak of 120 domestic manufacturers, 100 firms recently went out of business, taking with them 30,000 jobs.*

Ocean Thermal Energy Conversion: No systems in commercial operation.

Tidal Power: No systems in commercial operation.

Electricity from Industrial Processes' Waste Heat: From 1978–1988, some 2,800 of the "cogeneration" facilities enlarged U.S. generating capacity by about 4%*.

Wind Energy: Accounts for less than one-third of 1% of installed U.S. generating capacity.*

Electricity and Steam Heat from Geothermal Resources: Installed generating capacity is comparable to totals for wind, with most of it operating in a single California field.* Steam heat from geothermal sources is not a significant factor outside the scenic but constrained environs of Boise, Idaho.

Small-Scale Hydropower: We have about 1,500 of these units, which enlarge national generating capacity by about 1%.*

Wood-Fired Power Plants: U.S. utilities have built four of them. Total capacity is less than one-thirtieth that of the small-scale hydropower inventory.*

Solar Thermal Generators: We have eight systems, all of them in California, and their combined capacity does not even match that of the wood-fired units.*

Photovoltaics: The international industry's worldwide sales in 1986 were 23 megawatts, about half the capacity of the smallest U.S. wood-fired power plant.*

Alcohol-Based Fuels: Alcohol additives for gasoline exceeded 1.7 billion gallons in 1987, but still represented only about 1% of the energy content of the gasoline consumed during that year by U.S. vehicles. . . . *

Synthetic Fuels: Energy supply from synfuels plants is less than one-sixth the output of small hydroelectric units, notwithstanding continued operation of the mammoth Great Plains Coal Gassification Unit (which defaulted on more than $1.5 billion in federally guaranteed loans).*

Fusion Power: Technology still under development; no systems in or near commercial operation.

* Endnotes omitted.

Economic Costs

The costs and benefits of imposing carbon taxes are very difficult to assess. Some individuals feel that the economic costs of cutting back on greenhouse gas emissions are not great enough to justify actions such as a carbon tax. Here are the thoughts of Michael Boskin, chairman of former President George Bush's Council of Economic Advisors (taken from Bob Davis, "Bid to Slow Global Warming Could Cost U.S. $200 Billion a Year, Bush Aide Says," *Wall Street Journal,* April 16, 1990, p. B4):

> "The stakes are very high economically," said Michael Boskin, CEA chairman. "You'd likely wind up seeing a sharp reduction of economic growth" around the globe if the so-called greenhouse gases are to be reduced by 20% during the next 15 years. . . .
>
> Mr. Boskin . . . played down the significance of the potential heating. A rise of a few degrees Celsius would be similar to the change from moving from Boston to Washington, and might even benefit U.S. agriculture, he said.
>
> He also argued that people could use technology to adjust to the warmer weather. "For example, air conditioning has made previously less-hospitable climes much more hospitable," he said.
>
> Mr. Boskin said he and his staff hadn't done original research on the costs of greenhouse warming, but had reviewed a "substantial number" of U.S. computer models. Cutting emissions 20% by the year 2005, as has been suggested by countries in Western Europe, would cost the U.S. "trillions of dollars—$100 billion to $200 billion a year would be in the ballpark," he estimated.
>
> "It would mean a period of substantially higher unemployment and lower economic growth," he said, because the U.S. would have to switch to "much more expensive forms of energy."

Afforestation and Reforestation

There has been great dispute about the role of deforestation throughout the world as a factor in climate change. Many individuals argue that one way to combat global warming is to replant trees where they once grew or plant trees in new areas. In 1988, the American Forestry Association proposed a three-part plan to combat global warming along these lines. The following selection (taken from R. Neil Sampson, "ReLeaf for Global Warming," *American Forests,* November/December 1988, pp. 11–13) describes the third part of that proposal:

> Plant and care for the trees in their own yards and along their streets so that their dwelling becomes more pleasant, they use less energy,

and they can say they have contributed to the solution of a global problem. . . .

Our message will begin with an urban focus: 100 million "energy tree spaces" are available in urban communities, and filling those spaces could result in a savings of *40 billion kilowatt-hours* of energy and cut as much as 18 million tons of U.S. carbon dioxide production each year. Those trees would reduce the amount of carbon dioxide in the air—and the greenhouse effect—as much as a new forest larger than the state of Connecticut.

Compared to global emission levels, that amount is still small—a good deal less than one percent of the total. . . . But global issues need multiple approaches, and the right niche for any particular organization and citizen is something they *can do*, not just something they hope someone else will do. . . .

To further our goals, we've testified before Congress in support of the energy-conserving features of the National Energy Policy Act. . . . We've also asked Congress to amend that Act to create a new community-grant program that could be used for urban forestry projects deigned with energy-saving features.

We'll be going to Congress and to the U. S. Department of Agriculture to promote improved forest research, and improved forestry programs on both public and private lands. We'll work to get added tree-planting emphasis on marginal crop and pasture lands. . . .

Finally, we are asking Congress to stop underwriting any U.S.-aided projects that result in deforestation of tropical lands. . . .

Deforestation is a highly sensitive issue in many developing countries, however. Some of these nations claim that wood is a critical natural resource in their economy and that developed nations have no right telling developing nations how to use their resources. The following statement was made by Wen Lian Ting, Malaysia's leading representative to the Earth Summit, at a briefing just prior to the Rio meeting in June 1992 (item quoted below preceded by a single asterisk was taken from James Brooke, "Delegates from 4 Nations Warm to High-Profile Role: Global Powerbroker," *New York Times*, June 12, 1992, p. A10; item preceded by a double asterisk was taken from Elliot Diringer and Charles Petit, "U.S. Gets a Stern Warning on Eve of Rio Conference," *San Francisco Chronicle*, June 3, 1992, p. A2):

*The almost obsessional anxiety to have a forest convention is driven by concerns which have nothing to do with forests or trees. Developed countries wish to appease their public opinion and thus get electoral mileage out of forests.

Forests are clearly a sovereign resource—not like atmosphere and oceans, which are global commons. We cannot allow forests to be taken up in global forums.

**It boggles the mind that our forests have become pawns in the political chess games played in . . . the Northern Hemisphere. . . . [Malaysia is not going to keep its trees] in custody for those who have destroyed their own forests and now try to claim ours as part of the heritage of mankind.

5

Directory of Organizations

Nongovernmental Agencies

Center for Environmental Information
99 Court Street
Rochester, NY 14604
(716) 546-3796
FAX (716) 325-5131

The Center for Environmental Information is a private, nonprofit organization whose purpose is to collect and disseminate information on environmental issues, including climate change. The center maintains a library that is open to the general public and provides bibliographies and other information on environmental issues.

PUBLICATIONS: The center issues two monthly publications, *Global Climate Change Digest* and *Acid Precipitation Digest,* and a bimonthly newsletter for members, *CEI Sphere.*

Center for Global Change
University of Maryland
Executive Building, Suite 401
7100 Baltimore Avenue
College Park, MD 20740
(301) 454-0941
FAX (301) 454-0954

A division of the University of Maryland, the Center for Global Change coordinates research on climate change by university scientists, sponsors international conferences and symposia, and has compiled a catalogue of state and local laws dealing with global warming.

PUBLICATIONS: The center primarily publishes articles that result from local research on global warming and ozone depletion.

Climate Institute
316 Pennsylvania Avenue, SE
Suite 403
Washington, DC 20003
(202) 547-0104
FAX (202) 547-0111

The purpose of the Climate Institute is to increase public understanding about global warming and ozone depletion. The institute sponsors conferences and workshops and produces reports on climate change.

PUBLICATIONS: *Climate Alert,* a quarterly newsletter, as well as reports and monographs on institute conferences and workshops.

CONCERN
1794 Columbia Road, NW
Washington, DC 20009
(202) 328-8160

CONCERN was founded in 1970 to provide environmental information to individuals, community groups, educational institutions, public officials, and others involved with the environment, public education, and policy development. The organization's primary activity is the development, production, and distribution of community action guides dealing with topics such as global warming, pesticides, household wastes, farmlands, drinking water, and groundwater.

PUBLICATIONS: CONCERN has published seven guides on the topics listed above.

Electric Power Research Institute (EPRI)
3412 Hillview Avenue
P.O. Box 10412
Palo Alto, CA 94303
(415) 855-2272
FAX (415) 855-2954
Telex 82977 EPRI UF

The Electric Power Research Institute is a research organization funded through annual membership payments and project investments from more than 680 electric utilities in the United States. The goals of the institute are to discover, develop, and deliver advances in science and technology that will benefit member utilities, their customers, and the general society. To accomplish this goal the institute sponsors research and development projects that study the ways in which electricity is generated, delivered, and used, with special attention to the cost-effectiveness and environmental impact of these methods. The institute supports 31 research centers in 16 states.

PUBLICATIONS: Hundreds of technical reports have been produced as a result of research sponsored by EPRI. By far its most important and valuable publication for the general public, however, is its monthly *EPRI Journal.*

Environmental Action, Inc.
1525 New Hampshire Avenue, NW
Washington, DC 20036
(202) 745-4870

Environmental Action is a private, nonprofit, political action group whose purpose is to enhance the compatibility of humans and the environment by preventing pollution, reducing the use of nonrenewable resources, encouraging the conservation of energy and other natural resources, and eliminating the disruption of natural cycles. The organization carries out a number of activities, including helping individuals to organize at the grassroots level, making campaign contributions, offering information on environmental issues through a hotline and other methods, and lobbying Congress.

PUBLICATIONS: In addition to a variety of publications on environmental issues, Environmental Action publishes a bimonthly journal, *Environmental Action Magazine.*

Environmental Defense Fund
257 Park Avenue South
New York, NY 10010
(212) 505-2100
FAX (212) 505-2375

The goal of the Environmental Defense Fund is to develop and promote creative and economically viable solutions to environmental problems through a combination of scientific and legal means. The fund supports scientific research and public education on environmental issues. In addition it initiates and provides financial support for legal action needed to deal with certain environmental problems.

PUBLICATIONS: The fund has published a wide variety of books, pamphlets, articles, and papers on topics such as ozone depletion, acid rain, toxic materials, biotechnology, acid rain, wildlife conservation, and climate change. In addition it publishes a bimonthly newsletter, *EDF Letter.*

Friends of the Earth
530 Seventh Street, SE
Washington, DC 20003
(202) 544-2600

Friends of the Earth was founded in 1969 by David Brower, formerly executive director of the Sierra Club. The organization was established to work for the preservation, restoration, and rational use of the planet's natural resources. The organization supports scientific research and disseminates information to the general public on environmental issues. Friends of the Earth also works to introduce and influence legislation on environmental topics. A related group, the League of Conservation Voters, raises funds for congressional candidates who support its goals.

PUBLICATIONS: A list of books, pamphlets, and articles published by Friends of the Earth is available from the organization. In addition the group publishes a bimonthly newsletter, *Not Man Apart*.

Global Foundation, Inc.
P.O. Box 24-8103
Coral Gables, FL 33124-8103
(305) 284-4458

The Global Foundation is a private, nonprofit organization founded to support research and to disseminate information on environmental problems. The foundation sponsors lectures, workshops, and conferences of scientists, industrialists, government officials, and representatives from nongovernmental organizations. Its best known meeting is the annual International Scientific Forum on Energy.

PUBLICATIONS: The foundation does not publish any regular journal or newsletter but does produce books, research papers, news releases, and other reports from its seminars, workshops, and conferences.

Global ReLeaf Project
American Forestry Association
P.O. Box 2000
Washington, DC 20013
(202) 667-3300

The Global ReLeaf Project was created by the American Forestry Association in an effort to help combat global warming by preserving existing forests and planting new trees. One of the first programs sponsored by the project was an effort to plant 100 million trees in urban regions of the United States. In addition the project has developed a program of public education to help people understand the need for more and healthier trees and a program to influence national and international policies on the wise use of forests.

PUBLICATIONS: The project has published a booklet explaining how citizens can become involved in tree planting, *Global ReLeaf Action Guides,* and also publishes a regular newsletter, *Global ReLeaf Report.*

The Institute for Earth Education
Box 288
Warrenville, IL 60555

The Institute for Earth Education is an international, nonprofit, educational organization made up of a network of individuals and organizations committed to fostering Earth education programs. The institute has branches in the United States, Canada, Great Britain, France, Italy, Japan, Australia, and New Zealand. The group develops and disseminates educational programs designed to help people build an understanding for and appreciation of the harmony between human life and the environment. It conducts workshops, hosts international and regional conferences, supports local groups, distributes an annual catalog, and publishes books and program materials.

PUBLICATIONS: In addition to books and other educational materials for readers from elementary school through adulthood, the institute publishes a journal, *Talking Leaves.*

National Wildlife Federation
1400 16th Street, NW
Washington, DC 20036-2266
(703) 790-4000

The goal of the National Wildlife Federation is to inform the general public and legislative bodies about important environmental issues. It works actively through state affiliates and individual members to influence state and national conservation policies through administrative, legislative, and legal channels. It also maintains natural resource law and science centers throughout the United States. The federation also sponsors educational trips that focus on conservation issues, wildlife camps for children, leadership training programs, and an annual Washington Action Workshop.

PUBLICATIONS: Members receive one or more of the following magazines and newsletter: *EnviroAction, National Wildlife, International Wildlife, Ranger Rick,* and *Your Big Backyard.* In addition the federation has published a number of scientific, technical, and general interest books, pamphlets, and nonprint materials, such as the 11-page *Global Warming: A Personal Guide to Action* and the videotape and curriculum booklet *Our Threatened Heritage.*

Natural Resources Defense Council
1350 New York Avenue, NW
Washington, DC 20005
(202) 783-7800
FAX (202) 783-5917

The Natural Resources Defense Council was founded to protect the nation's air, water, and food supplies through legal action, scientific research, advocacy, and public education. One of the council's most effective approaches has been to sue private companies that pollute the environment. Recently it has expanded its efforts to the international level, promoting exchanges between individual scientists and scientific organizations from the United States, the former Soviet Union, and other nations.

PUBLICATIONS: Members of the Natural Resources Defense Council receive *The Amicus Journal* four times a year and *NRDC Newsline* five times a year. In addition the council has published a number of books and pamphlets dealing with environmental issues, including global warming. Among these are *The Statehouse Effect: State Policies To Cool the Greenhouse*, by Daniel A. Lashof and Eric A. Washburn; *Cooling the Greenhouse: Vital First Steps To Combat Global Warming; Farming in the Greenhouse: What Global Warming Means for American Agriculture*, by Justin R. Ward, Richard A. Hardt, and Thomas E. Kuhnle; and *The Rising Tide: Global Warming and World Sea Levels*, by Lynne Edgerton.

North American Association for Environmental Education
P.O. Box 400
Troy, OH 45373
(513) 339-6835

The North American Association for Environmental Education is a professional association of educators and administrators interested or involved in environmental education at all grade levels. Its purpose is to assist and support the activities of its members. The association holds an annual conference in a different part of the nation each year.

PUBLICATIONS: The association publishes the bimonthly newsletter *Environmental Communicator*.

Renew America
1400 16th Street, NW
Suite 710
Washington, DC 20036
(202) 232-2252
FAX (202) 232-2617

Renew America is a national clearinghouse for environmental information. The organization's annual environmental awards program identifies, reviews, and promotes successful environmental programs from across the United States. The programs that are selected are chosen because they deal with one or more of the twenty most pressing environmental problems facing the nation today. The "success stories" are listed in the organization's annual *Environmental Success Index*. By bringing attention to these programs, Renew America attempts to provide examples of solutions that people can apply on a local, regional, or national level.

PUBLICATIONS: In addition to the *Environmental Success Index,* Renew America has published books such as *Sustainable Energy, State of the States Reports* (annual), *The Oil Rollercoaster: A Call to Action,* and *Reducing the Rate of Global Warming: The States' Role.*

Resources for the Future
1616 P Street
Washington, DC 20036
(202) 328-5006

Resources for the Future is an independent, nonprofit organization originally established through a grant from the Ford Foundation. The purpose of the organization is to inform and improve policy debates about issues relating to natural resources and the environment. It sponsors research; carries out educational programs; holds conferences and other types of meetings; provides grants and fellowships; and arranges policy briefings for legislators, business leaders, and the media. The organization does not, itself, take a stand on any side of any particular issue, but maintains a nonpartisan position in all of its work.

PUBLICATIONS: In addition to more than 100 books and reports, Resources for the Future publishes the useful quarterly newsletter *Resources,* which contains news of research and policy studies on issues dealing with the environment and natural resources. The newsletter is free to the public.

The Sierra Club
730 Polk Street
San Francisco, CA 94109
(415) 776-2211

The Sierra Club is one of the oldest environmental organizations in the world. It was founded on May 28, 1892, by a group of Californians interested in protecting the natural environment. The club's first president was John Muir. Today the club has more than 600,000 members in 58 chapters and 403 subgroups. Its purpose is "to explore, enjoy, and protect the wild places of the earth; to practice and promote the responsible use of the earth's ecosystems and resources; to educate and enlist humanity to protect and restore the quality of the natural and human environment; and to use all lawful means to carry out these objectives."

PUBLICATIONS: Members of the Sierra Club receive its official magazine, *Sierra,* six times a year and its *National News Report* 24 times a year. In addition the club publishes a wide variety of books, kits, teaching units, slide shows, films, filmstrips, videos, reports, pamphlets, and other materials. Among these are "The Tropical Rainforest" (slide show), "Global Warming Activist Video," "Global Warming" (factsheet), "21 Ways To Help Stop Global Warming by the 21st Century" (factsheet), *Global Warming* (book), and "Climate in Crisis" (poster).

Union of Concerned Scientists
26 Church Street
Cambridge, MA 02138
(617) 547-5552

The Union of Concerned Scientists is a nonprofit organization of nearly 100,000 scientists and other interested citizens concerned about the impact of advanced technology on society. Its programs focus on national energy policy, arms control, and nuclear power safety. In addition to producing books, reports, pamphlets, brochures, briefing papers, and visual aids on these topics, the union provides speakers for civic groups, educational programs, and professional associations. The organization also works actively to influence legislative action on the topics in which it is interested.

PUBLICATIONS: Among the union's many publications are a video, "Greenhouse Crisis: The American Response"; a report, *Cool Energy: The Renewable Solution to Global Warming;* and two brochures, "The Global Warming Debate: Answers to Controversial Questions" and "How You Can Fight Global Warming: An Action Guide."

World Resources Institute
1709 New York Avenue, NW
Washington, DC 20006
(202) 638-6300

The World Resources Institute is an independent, nonprofit organization that receives financial support from private foundations, governmental institutions, private corporations, and interested individuals. The work of the institute is carried out by a staff of about 100 scholars, primarily in the field of science and economics. When the need arises, outside consultants are hired to assist with projects. The purpose of the institute is to help governments, the private sector, environmental and development organizations, and other groups to find ways to encourage economic growth without causing damage to the environment. In addition to carrying out research, the institute publishes books, papers, and reports; holds briefings, seminars, and conferences; and provides the print and nonprint media with background information.

PUBLICATIONS: In 1992, the institute advertised 87 different publications, ranging from large database and reference books such as *World Resources, 1992–93* and *Global Biodiversity 1992: Status of Earth's Living Resources* to shorter paperback books such as *Greenhouse Warming: Negotiating a Global Regime* and *The Greenhouse Trap: What We're Doing to the Atmosphere and How We Can Slow Global Warming.*

Worldwatch Institute
1776 Massachusetts Avenue, NW
Washington, DC 20036-1904
(202) 452-1999

The Worldwatch Institute was founded in 1974 to provide policymakers and the general public with information about links between the world economy and the environment. The institute attempts to raise public awareness about environmental issues to the point where some kind of policy response is indicated. Researchers at the institute carry out their own fieldwork and review research conducted by other scientists throughout the world. They convey information to government officials, journalists, academics, and the general public. Institute officials give interviews to all forms of the media, including U.S. and international newspapers, radio, and television stations and networks.

PUBLICATIONS: The institute publishes the annual book *State of the World,* in which it provides an overall review of the state of the world's environment. Its other publications include *Worldwatch Magazine, The Worldwatch Reader on Global Environmental Issues,* and more than 100 *Worldwatch Papers* on topics such as global warming, biological diversity, poverty and the environment, the "green" revolution, reforesting the Earth, and ozone depletion.

Governmental Agencies

Carbon Dioxide Information Analysis Center
MS-6335, Building 1000
Oak Ridge National Laboratory
P.O. Box 2008
Oak Ridge, TN 37831-6335
(615) 574-0390
FAX (615) 574-2232

The Carbon Dioxide Information Analysis Center (CDIAC) is part of the Environmental Sciences Division of the Oak Ridge National Laboratory. It was created in 1982 to provide information about the connection between elevated levels of carbon dioxide in the atmosphere and global change, information that can be used in research, policymaking, and education on both national and international levels. CDIAC activities include (1) the collection and evaluation of data, articles, and reports on carbon dioxide and other greenhouse gases and the environment and (2) the production of print and nonprint data packages summarizing the results of this research.

PUBLICATIONS: In addition to a large number of individual reports and data packages, CDIAC also has published *TRENDS '91: A Compendium of Data on Global Change, Catalog of Data Bases and Reports, and Glossary: Carbon Dioxide and Climate.* It also publishes the regular newsletter *DOE Research Summary.*

Climate Program Office and Atmospheric Environment Service
Environment Canada
4905 Dufferin Street
Downsview, ON M3H 5T4
Canada
(416) 739-4763

The Climate Program Office was initiated by Canada's department in charge of environmental affairs, Environment Canada. It was established to carry out research on the social and economic impacts of a warmer climate on Canada. The program combines the expertise of experts from government, industry, and higher education. The Atmospheric Environment Service is a division of Environment Canada which publishes a number of valuable documents concerning climate change and its impact on Canada.

PUBLICATIONS: The Atmospheric Environment Service publishes an ongoing series of reports, *Climate Change Digest,* which includes titles such as *Global Warming: Implications for Canadian Policy, Climate Change and Canadian Impacts: The Scientific Perspective,* and *Climate Warming and Canada's Comparative Position in Agriculture. CO_2/Climate Report* also is available from the office.

National Center for Atmospheric Research
P.O. Box 3000
Boulder, CO 80307-3000
(303) 497-1174

The National Center for Atmospheric Research is an important center for research by the National Oceanic and Atmospheric Administration. Please see the listing for the administration for more information.

UN Intergovernmental Panel on Climate Change
Case Postale N° 2300
1211 - Geneva 2
Switzerland
41 22 7308 215/254/284
FAX 41 22 733 1270
Telex: 41 41 990 MM CH

Under the auspices of the United Nations Environment Programme and the World Meteorological Organization, the UN Intergovernmental Panel on Climate Change (IPCC) was created in November 1988. The primary function of IPCC is to act as a central clearinghouse for the collection of research and dissemination of information about global warming.

PUBLICATIONS: As of late 1992, IPCC had released 13 publications, almost all of which deal with fairly technical information about global warming. The most important of these publications are those that make up the *IPCC First Assessment Report, 1990,* and the *1992 IPCC Supplement.*

U.S. Environmental Protection Agency
401 M Street, SW
Washington, DC 20460
(202) 260-2080

The U.S. Environmental Protection Agency was created by an act of Congress in 1970 to coordinate efforts of the federal government in protecting and enhancing the environment. It carries out activities to control and reduce pollution in the areas of air, water, solid waste, pesticides, radiation, and toxic substances. It sponsors and coordinates research and antipollution activities by federal, state, and local groups; public and private organizations; and individuals. It is responsible for setting standards, monitoring the environment, and enforcing laws and regulations dealing with the environment.

PUBLICATIONS: In addition to issuing many reports on environmental issues, the agency also publishes the monthly newsletter *EPA Journal.*

U.S. National Oceanic and Atmospheric Administration
Office of Global Programs
1100 Wayne Avenue, Suite 1225
Silver Springs, MD 20910
(301) 427-2089
FAX (301) 427-2082
FAX (301) 427-2073

Research on global warming is scattered throughout dozens of agencies within the U.S. government. In addition to the National Oceanic and Atmospheric Administration (NOAA), the National Aeronautics and Space Administration (NASA), the National Science Foundation (NSF), and the Environmental Protection Agency (EPA) have been the major sponsors of scientific research on climate change. Since 1989, NOAA has developed an extensive research program that uses Earth- and satellite-based systems to monitor changes in oceanic and atmospheric properties. Some of the most important work being done by NOAA scientists is located at the National Center for Atmospheric Research, to which you should also refer.

PUBLICATIONS: NOAA publishes a large number of booklets, pamphlets, and other materials dealing with its global change research. Some examples are *TOGA: Tropical Ocean Global Atmosphere Program, Atlantic Climate Change Program Science Plan, Product Development Plans for Operational Satellite Products,* and *Global Precipitation Science Plan for the NOAA Climate and Global Change Program.* Its more general titles include *Reports to the Nation on Our Changing Planet, The Vision: A Rededication of NOAA,* and *Our Changing Planet: U.S. Global Change Research Program* (published annually describing funding and programs for each fiscal year).

U.S. National Science Foundation
Committee on Earth and Environmental Sciences
1800 G Street, NW, Room 232
Washington, DC 20550
(202) 357-7861

Research on global warming and climate change in the U.S. government is coordinated by the U.S. Global Change Research Program (USGCRP), a program developed by the interagency Committee on Earth and Environmental Sciences. Since fiscal year 1990, USGCRP has reflected the nation's priorities in developing research designed to resolve uncertainties in human knowledge about natural and anthropogenic changes now taking place in the Earth's environment.

PUBLICATIONS: The committee's annual report is *Our Changing Planet,* a document that outlines the research emphases of the U.S. government for the coming year.

State Programs

Many colleges, universities, and research centers have special divisions or programs to study climate change. The following are representative of those found throughout the nation.

California
Climate Research Group
Scripps Institute of Oceanography
University of California at San Diego - A024
La Jolla, CA 90093

Colorado
Cooperative Institute for Research in the Atmosphere
Colorado State University
Fort Collins, CO 80523

Delaware
Center for Climatic Research
Department of Geography
University of Delaware
Newark, DE 19716

Hawaii
Joint Institute for Marine and Atmospheric Research
University of Hawaii
1000 Pope Road, MSB 312
Honolulu, HI 96822

Nevada
Atmospheric Sciences Center
P.O. Box 60220
University of Nevada
Reno, NV 89506

Ohio
Atmospheric Sciences Program
Ohio State University
2015 Neil Avenue
Columbus, OH 43210

Oregon
Climatic Research Institute
Oregon State University
811 S.W. Jefferson Street
Corvallis, OR 87330

Rhode Island
Center for Atmospheric Chemistry Studies
University of Rhode Island
Narragansett, RI 02882

Washington
Joint Institute for the Study of the Atmosphere and Ocean
AK-40, 606 Atmospheric Sciences Building
University of Washington
Seattle, WA 98195

Wisconsin
Center for Climatic Research
University of Wisconsin
1225 West Dayton Street
Madison, WI 53706

6

Selected Print Resources

Bibliography

The Greenhouse Effect: A Bibliography. Santa Cruz, CA: Reference and Research Services, 1990. 60p. ISBN 0-937855-34-0.

Although slightly out of date, this bibliography is far and away the most valuable single reference on books and articles dealing with every aspect of global warming and ozone depletion. It contains separate sections on general resources, greenhouse gases, CFCs, energy resources, forests, natural resources, water resources, agriculture, international aspects, and possible solutions.

Books

Abrahamson, Dean Edwin, ed. **The Challenge of Global Warming.** Washington, DC: Island Press, 1989. 356p. ISBN 0-933280-87-4.

A collection of articles, speeches, and reports dealing with global warming. The book is divided into five parts: "The Challenge of Global Warming," "Global Warming: Biotic Systems," "Global Warming: Physical Impacts," "The Greenhouse Gases," and "Policy Responses." The book is a product of the Natural Resources Defense Council, which in 1988 launched the Atmospheric Protection Initiative, a campaign to develop solutions to the worldwide crisis of atmospheric pollution.

Acid Rain Foundation. **Information Packet/Global Climate Change.** Raleigh, NC: Acid Rain Foundation, 1992.

A collection of brochures, booklets, legislation, and articles published by scientists, industry, government, environmental, and public interest groups. The packet is recommended for use in debates, research papers, reports, and vertical files.

Balling, Robert C., Jr. **The Heated Debate: Greenhouse Predictions versus Climate Reality.** San Francisco: Pacific Research Institute for Public Policy, 1992. 195p. ISBN 0-936488-47-6.

Balling is one of the leading spokespersons for the position that the problem of global warming has been greatly exaggerated. He attempts to demonstrate that the scientific information needed to support the global warming position is either very weak or nonexistent. He urges readers to ignore emotional appeals about the issue and begin thinking more rationally about the status of the Earth's climate.

Barth, Michael C., and James G. Titus. **Greenhouse Effect and Sea Level Rise: A Challenge for This Generation.** New York: Van Nostrand Reinhold, 1984. 325p. ISBN 0-442-20991-6.

What changes can be expected if global warming should result in a significant rise in the world's sea level? This book examines this question from a number of perspectives. It opens with a general review of the factors that affect sea level, with special attention to the effects of changing climate and some estimates of future sea-level rise. Then a series of authors describe specific effects that might be expected at Charleston, South Carolina, Galveston Bay, Texas, and other locations. Possibly the best single resource on this topic.

Bates, Albert K. **Climate in Crisis: The Greenhouse Effect and What We Can Do.** Summertown, TN: Book Publishing, 1990. 230p. ISBN 0-913990-67-1.

This book for young adults and adult readers presents a worst-case scenario that ignores much of the doubt that remains about climate change. Those who believe global warming is a proved phenomenon will be pleased to see their case stated so strongly, but those with doubts about it will ask where the opposite side is to be found. The author suggests a number of actions that can be taken to deal with global warming, although not all suggestions will be endorsed by everyone interested in the topic.

Blashfield, Jean F., and Wallace B. Black. **Global Warming.** Chicago: Children's Press, 1991. 126p. ISBN 0-516-05501-1.

The authors have written a charming book that clearly explains the phenomena of global warming and ozone depletion for young readers. They present the case for climate change, but point out that scientific controversy still continues as to the extent of that phenomenon. The book opens with a description of what life would be like in the year 2044 if global warming takes place according to some predictions now being made by scientists. The authors also provide some simple experiments to illustrate some of the fundamental points about the greenhouse effect and conclude with recommendations for individual action to deal with these two environmental problems.

Bolin, Bert, Bo R. Döös, Jill Jäger, and Richard A. Warrick. **The Greenhouse Effect, Climatic Change, and Ecosystems.** New York: John Wiley, 1986. 541p. ISBN 0-471-91012-0.

A publication of the Scientific Committee on Problems of the Environment (SCOPE), a committee established by the International Council of Scientific Unions. The book is a collection of scientific reports covering all aspects of global change, including the release of carbon dioxide to the atmosphere, the emission of other greenhouse gases, estimates of climate change based on increasing levels of carbon dioxide, predictions of sea-level change, and the impact of changing climate on terrestrial systems. The extensive bibliographies at the conclusion of each section are especially useful.

Brower, Michael. **Cool Energy: The Renewable Solution to Global Warming.** Cambridge, MA: Union of Concerned Scientists, 1990. 89p. ISBN 0-938987-12-7.

This report discusses the status of various renewable-energy technologies, their near- and long-term prospects, and their potential role in slowing global warming. It also proposes some cost-effective policies to accelerate the development of renewable energy sources.

Brown, Lester R., et al. **State of the World.** New York: W. W. Norton, published annually. ISBN 0-393-02934-4 (1992 edition).

Perhaps the most comprehensive assessment of environmental issues throughout the world, this book is published annually with up-to-date information on topics such as ozone depletion, population issues, biodiversity, climate change, pollution, and other important topics.

Budyko, M. I. **Climate Changes.** Washington, DC: American Geophysical Union, 1977. 261p. ISBN 0-87590-206-5.

Budyko is one of the pioneers of climate modeling. He has had a significant influence on many of the climate modelers working in the

United States today. This book provides a classic introduction to the issue of climate modeling. Budyko writes about the origins of climate, present climatic changes, climates of the past, the relationship of climate to evolution, human effects on the climate, and climates of the future.

————. **The Earth's Climate.** New York: Academic Press, 1982. 307p. ISBN 0-12-139460-3.

This book is largely an updating of *Climate Changes*, listed immediately above.

Byrne, Jeanne. **Global Warming: A Personal Guide to Action** (photocopied). New York: National Wildlife Federation, n.d. 11p. Item #79978.

An overall review of the global warming issue with suggestions for changes in personal lifestyle that can help slow the increase in atmospheric temperatures.

Carbon Dioxide Research Division. **Atmospheric Carbon Dioxide and the Global Carbon Cycle.** Washington, DC: U.S. Department of Energy, December 1985. DOE/ER-0239. Edited by John R. Trabalka.

————. **Characterization of Information Requirements for Studies of CO_2 Effects: Water Resources, Agriculture, Fisheries, Forests and Human Health.** Washington, DC: U.S. Department of Energy, December 1985. DOE/ER-0236. Edited by Margaret R. White.

————. **Detecting the Climatic Effects of Increasing Carbon Dioxide.** Washington, DC: U.S. Department of Energy, December 1985. DOE/ER-0235. Edited by Michael C. MacCracken and Frederick M. Luther.

————. **Direct Effects of Increasing Carbon Dioxide on Vegetation.** Washington, DC: U.S. Department of Energy, December 1985. DOE/ER-0238. Edited by Boyd R. Strain and Jennifer D. Cure.

————. **Glaciers, Ice Sheets, and Sea Level: Effect of a CO_2-Induced Climatic Change.** Washington, DC: U.S. Department of Energy, December 1985. DOE/ER/60235-1.

————. **Projecting the Climatic Effects of Increasing Carbon Dioxide.** Washington, DC: U.S. Department of Energy, December 1985. DOE/ER-0237. Edited by Michael C. MacCracken and Frederick M. Luther.

The preceding six reports were written to assess the changes in knowledge that had come about since a 1977 meeting of scientists in Miami dealing with the carbon cycle and the effects on that cycle produced by the release of anthropogenic carbon dioxide. The reports contain some of the best technical information available on each of the topics described by the report titles.

CONCERN. **Global Warming & Energy Choices.** Washington, DC: CONCERN, 1991. 38p. ISBN 0-937345-07-5.

This paperback booklet is the eighth in CONCERN's series of community action guides, designed to raise public awareness of the implications of energy choices made in the United States. It outlines methods by which citizens can make choices in their own lives and push for community action that will have effects on global warming.

The Crucial Decade: The 1990s and the Global Environmental Challenge. Washington, DC: World Resources Institute, 1989. 30p. ISBN 0-915825-37-6.

This book brings together brief descriptions of critical world environmental problems and provides a checklist of measures to deal with them. It focuses particularly on large-scale atmospheric changes and loss of biological diversity. The book makes the argument that concerted efforts can be made to deal with these problems. It also outlines the role of population in the origin and solution of the problems.

Dunnette, David A., and Robert J. O'Brien, eds. **The Science of Global Change: The Impact of Human Activity on the Environment.** Washington, DC: American Chemical Society, 1992. 506p. ISBN 0-8412-2197-9.

Written for the nonspecialist, this book examines the interaction among the hydrosphere, atmosphere, lithosphere, and biosphere, with special attention to the effects of human activities on these global elements.

Edgerton, Lynne T. **The Rising Tide: Global Warming and World Sea Levels.** Washington, DC: Island Press, 1991. 140p. ISBN 1-55963-068-X.

The author of this book is an attorney with the Natural Resources Defense Council (NRDC). NRDC tends to accept some of the more pessimistic predictions for sea-level rise as a result of global warming. This book focuses, therefore, on some of the very practical problems coastal areas may face as a result of higher sea levels. These problems include the transfer of hazardous waste, flooding of low-lying areas, and the relocation of communities. Appendixes provide a valuable list of government documents relating to this problem.

Ephron, Larry. **The End: The Imminent Ice Age and How We Can Stop It.** Berkeley, CA: Celestial Arts, 1988. 233p. ISBN 0-89087-507-3.

Ephron presents an intriguingly different angle on the debate about global warming. He summarizes a theory that glacial periods can be brought on when the greenhouse effect gets out of control, a condition that he claims now exists on the Earth. The long-term effect of increasing concentrations of carbon dioxide and other greenhouse gases in the atmosphere will not be a warmer climate, he says, but a new ice age. He then explains what we should expect in such a period and outlines some ways that humans can act to prevent another glacial era. The book has been the subject of considerable debate within the scientific community.

Firor, John. **The Changing Atmosphere: A Global Challenge.** New Haven, CT: Yale University Press, 1990. 145p. ISBN 0-300-03381-8.

Firor shows how a number of environmental problems, such as acid rain, ozone depletion, and global warming, are interrelated.

Further, he argues that an important factor in the worsening of these problems is out-of-control human population growth. He suggests that humans soon will have to make a decision as to whether they are willing to live in a degraded environment that will inevitably result from these problems or whether they will take positive actions to reduce the severity of these phenomena. More aggressive programs of population control would be one such action, he says.

Fisher, David E. **Fire and Ice: The Greenhouse Effect, Ozone Depletion, and Nuclear Winter.** New York: Harper & Row, 1990. 232p. ISBN 0-06-016214-7.

Fisher discusses these three related topics in two separate parts. In the first part he outlines the scientific background for each of the issues and shows how they are related to each other. In the second part of the book he reviews some possible ways to deal with each problem.

Fishman, Jack, and Robert Kalish. **Global Alert: The Ozone Pollution Crisis.** New York: Plenum Press, 1990. 311p. ISBN 0-306-43455-5.

Although somewhat technical and detailed, this book is likely to appeal to many readers. Fishman has been involved in ozone research at NASA's Langley Research Center, and he argues that the build up of ozone in the troposphere may be more of a problem than is the depletion of that gas in the stratosphere. The authors show how three apparently unrelated problems—global warming, tropospheric ozone, and stratospheric ozone—are actually connected to each other.

Flavin, Christopher. **Slowing Global Warming: A Worldwide Strategy.** Washington, DC: Worldwatch Institute, 1989. 94p. ISBN 0-916468-92-5.

One of a series of Worldwatch Papers (#91) dealing with global problems. The emphasis in this book is on possible ways of dealing with the emission of greenhouse gases. It contains a good section on efforts being made by other nations to deal with the problem of global warming.

Gribbin, John. **Hothouse Earth: The Greenhouse Effect and Gaia.** New York: Grove Weidenfeld, 1990. 273p. ISBN 0-8021-1374-5.

This superb book by an outstanding science writer supplies a complete and understandable description of the science behind the greenhouse effect and the potential impact the greenhouse effect may have on the environment. An especially interesting approach is the author's use of Jim Lovelock's concept of the Earth as "Gaia" to discuss this subject. It's easy to be overwhelmed by the mass of facts that Gribbin presents, but it's worth staying with the sometimes dense prose to follow the fascinating story he conveys.

Gribbin, John, ed. **Climatic Change.** Cambridge: Cambridge University Press, 1986. 280p. ISBN 0-521-21594-3.

This fairly technical collection of articles provides an excellent overview of the question of global warming. Individual papers deal with climates of the past, the Earth's "heat budget," astronomical influences on the Earth's climate, modeling climate change, and potential effects of climate change on human societies. Many of the premier climate researchers in the world (Budyko, Kellogg, and Schneider, for example) contributed chapters to this book.

Gribbin, John, and Mick Kelly. **Winds of Change.** London: Headway, 1989. 162p. ISBN 0-340-51505-8.

Gribbin's *Hothouse Earth* tends to concentrate on the scientific aspects of global warming. In *Winds of Change* he writes on the *assumption* that global warming is a reality and then outlines the social, political, economic, and other nonscientific implications that result from this assumption. He uses the U.K. Meteorological Office projections as the basis for his analysis.

Hare, Tony. **The Greenhouse Effect.** New York: Franklin Watts, 1990. 32p. ISBN 0-531-17217-1.

This short book is nicely illustrated and organized so as to present specific topics in two-page spreads. The treatment is complete, with almost no conceivable factor ignored. Industrial emissions, burning of forests, sources and effects of greenhouse gases other than carbon dioxide, and ozone depletion all are given attention. In such a short book, of course, the treatment of these topics is brief. Diagrams are well used in most cases. The book is appropriate for readers in grades 5 to 8.

Hocking, Colin, Cary Sneider, John Erickson, and Richard Golden. **Global Warming.** Berkeley, CA: Lawrence Hall of Science, 1990. 171p. ISBN 0-912511-75-3.

The authors attempt to present a balanced discussion about the possibility of global warming and to describe a number of actions individuals and society can take to deal with this problem. The book is written for students in grades 7 through 10.

Idso, Sherwood B. **Carbon Dioxide: Friend or Foe?** Tempe, AZ: IBR Press, 1982. 92p. ISBN 0-9623489-0-2.

"This volume will be controversial." Idso opens his book with this statement, pointing out that he intends to show that the dire effects of global warming that so many scientists predict will not really occur. In fact, he tries to show that there will be "minimal climate (temperature) change from a doubling of atmospheric CO_2." Idso argues that the role of plants in absorbing carbon dioxide has been ignored in previous research.

————. **Carbon Dioxide and Global Change: Earth in Transition.** Tempe, AZ: IBR Press, 1989. 292p. ISBN 0-9623489-1-0.

Idso updates his arguments about the controversy over global warming first enunciated in *Carbon Dioxide: Friend or Foe?* (see above).

Imbrie, John, and Katherine Palmer Imbrie. **Ice Ages: Solving the Mystery.** New York: Enslow, 1978. 224p. ISBN 0-89490-015-3.

A very interesting discussion of astronomical theories about the origin of the ice ages with a thorough discussion of the life and work of Milutin Milankovitch. The first author was involved in some of the later research on this topic. The book provides a good background on the astronomical aspects of this topic but contains no discussion of the role of carbon dioxide and other greenhouse gases in global warming.

Jäger, J., and H. L. Ferguson. **Climate Change: Science, Impacts and Policy.** Cambridge: Cambridge University Press, 1990. 578p. ISBN 0-521-41631-0.

This volume contains the proceedings of the 2nd World Climate Conference, held in Geneva in 1990. The book is an invaluable source of information on nearly every aspect of the global warming question. It contains reports of the most recent research on climate change as well as research on the potential effects of such change on agriculture, water resources, energy use, oceans and fisheries, the environment, and human societies.

Johnson, Rebecca L. **The Greenhouse Effect: Life on a Warmer Planet.** Minneapolis, MN: Lerner, 1990. 112p. ISBN 0-8225-1591-1.

This book has been highly praised because of its well-balanced, clear presentation of the problem of climate change. One reviewer has called it an example of "the very best work that can be done in the production of excellent science books for children." The author provides the usual description of how the greenhouse effect occurs and how human activities alter that natural process. She also focuses on the potential effects global warming may have on the natural and human environment of the twenty-first century.

Kellogg, William, and Robert Schware. **Climate Change and Society: Consequences of Increasing Atmospheric Carbon Dioxide.** Boulder, CO: Westview, 1981. 178p. ISBN 0-86531-179-X.

A book that is now somewhat dated, this title is important as an early effort to try to outline the kinds of changes that societies will have to make if global warming becomes a reality. The references cited provide a comprehensive review of earlier works dealing with the topic of climate change.

Liss, P. S., and A. J. Crane. **Man-Made Carbon Dioxide and Climate Change: A Review of Scientific Problems.** Norwich, England: Geo Books, 1983. 127p. ISBN 0-86094-140-X.

Produced originally as a technical report for the Central Electricity Generating Board in London, this book examines (1) the global behavior of anthropogenic carbon dioxide and (2) the climatic consequences of increased levels of atmospheric CO_2.

Lovelock, James. **The Ages of Gaia: Biography of Our Living Earth.** New York: W. W. Norton, 1988. 252p. ISBN 0-393-02583-7.

Written a decade after the Gaia hypothesis was first developed, this book provides a more complete and more up-to-date discussion of the theory than did Lovelock's original book, *Gaia*, published in 1979. By 1988, the Gaia hypothesis had been attacked and defended by scholars from many different fields. This book reviews those arguments and makes a new defense of the idea.

Lyman, Francesca. **The Greenhouse Trap: What We're Doing to the Atmosphere and How We Can Slow Global Warming.** Boston: Beacon Press, 1990. 190p. ISBN 0-8070-8502-2.

A publication of the World Resources Institute that provides a good history of the global warming problem and describes ways that societies and individuals can act to reduce the release of greenhouse gases. A good selection of organizations and books that individuals can use is included.

MacCracken, Michael C., and Frederick M. Luther. **Detecting the Climatic Effects of Increasing Carbon Dioxide.** See Carbon Dioxide Research Division.

————. **Projecting the Climatic Effects of Increasing Carbon Dioxide.** See Carbon Dioxide Research Division.

Mackenzie, James J. **Breathing Easier: Taking Action on Climate Change, Air Pollution, and Energy Insecurity.** Washington, DC: World Resources Institute, 1988. 24p. ISBN 0-915825-35-X.

The author tries to show how the combination of climate change, persistent air pollution, and growing dependence on imported oil result in serious economic, social, and scientific problems for the United States. He outlines technological and policy actions that can help resolve these problems. This paperback booklet is beautifully illustrated and contains many clear and helpful charts.

McKibben, Bill. **The End of Nature.** New York: Random House, 1989. 226p. ISBN 0-394-57601-2.

This very well written book not only describes the facts about global warming but also attempts to analyze the philosophical significance of these changes for the place of humankind in nature. A highly recommended general introduction to the subject.

Malone, T. F., and J. G. Roederer, eds. **Global Change.** Cambridge: Cambridge University Press, 1984. 512p. ISBN 0-521-30670-1.

This book contains the proceedings of a symposium sponsored by the International Council of Scientific Unions in Ottawa, Canada, on September 25, 1984. An important focus of the symposium was the plans then being made for the International Geosphere-Biosphere Program. A number of papers deal specifically with that program and describe activities planned for further study of the global warming problem. Some of the papers are fairly technical, but many are accessible to the average reader.

Mathews, Jessica Tuchman, ed. **Greenhouse Warming: Negotiating a Global Regime.** Washington, DC: World Resources Institute, 1991. 80p. ISBN 0-915825-70-8.

A group of writers investigate the question of what kind of worldwide political organization can be developed to deal with the issue of global warming.

Mintzer, Irving. **A Matter of Degrees: The Potential for Limiting the Greenhouse Effect.** Washington, DC: World Resources Institute, 1987. 72p. ISBN 0-915825-27-9.

The author has developed a computer model that projects future emissions for six greenhouse gases that contribute most to global warming. He recommends policy interventions that could be adopted to control the effect of greenhouse gases on future climate change.

Mitchell, George J. **World on Fire: Saving an Endangered Earth.** New York: Charles Scribner's Sons, 1991. 247p. ISBN 0-684-19231-4.

Written by the majority leader of the U.S. Senate, this book concentrates on the kinds of political actions that can be taken domestically and internationally to deal with the problem of global warming.

Myers, Norman. **Not Far Afield: U.S. Interests and the Global Environment.** Washington, DC: World Resources Institute, 1987. 84p. ISBN 0-915825-24-4.

This book provides a general overview of worldwide environmental problems. The author reviews issues such as pollution across national boundaries, deforestation, depletion of water resources, and pressure on natural resources. He then outlines ways in which the United States can act to preserve global resources.

National Research Council, Carbon Dioxide Assessment Committee. **Changing Climate: Report of the Carbon Dioxide Assessment Committee.** Washington, DC: National Academy Press, 1983. 496p. ISBN 0-309-03425-6.

A series of papers prepared for the National Research Council dealing with topics such as the future of carbon dioxide emissions from fossil fuels, past and future atmospheric concentrations of carbon dioxide, detection and monitoring of carbon dioxide–induced climate changes, agricultural and biotic effects of carbon dioxide, the oceans as a sink for carbon dioxide, the role of methane in global warming, water supplies in a warmer world, and probable changes in sea levels as a result of global warming.

Ohio Sea Grant College Program. **Global Change Scenarios for the Great Lakes Region.** Columbus: Ohio Sea Grant College Program, 1992. Ten scenarios, 2–4p. each. OHSU-EP-07892.

The Ohio Sea Grant College Program has developed a set of ten scenarios that allow readers to examine possible futures for the Great Lakes region in a number of specific areas, including water resources, biological diversity, agriculture, recreation, forests, fish populations, shipping, and atmospheric change. The scenarios are designed for use in science classrooms.

Oppenheimer, Michael, and Robert H. Boyle. **Dead Heat: The Race against the Greenhouse Effect.** New York: Basic Books, 1990. 268p. ISBN 0-465-09804-5.

This exceptionally readable review of the global warming problem contains one of the most extensive lists of references to be found in any popular book on the subject.

Research Directorate of the National Defense University. **Climate Change to the Year 2000.** Washington, DC: Fort Lesley J. McNair, 1978. 109p. No ISBN.

This volume summarizes the opinions of climate research experts throughout the United States on a number of questions. Chapter 1 focuses on methods used to study climate. Chapter 2 outlines the variety of opinions about the extent of global warming that is to be expected, ranging from "large global cooling" to "large global warming." Chapter 3 summarizes the future climate scenarios provided by experts, concentrating especially on global temperature changes and changes in growing season and precipitation.

Revkin, Andrew. **Global Warming: Understanding the Forecast.** New York: Abbeville Press, 1992. 180p. ISBN 1-55859-310-1.

This richly illustrated book was developed as a companion publication for the special traveling exhibit on climate change developed by the American Museum of Natural History and the Environmental Defense Fund. The author describes those factors that have contributed to global warming and outlines a number of actions individuals can take to bring about positive change.

Rosenberg, N. J., W. E. Easterling, III, P. Crosson, and J. Darmstadter, eds. **Greenhouse Warming: Abatement and Adaptation.** Washington, DC: Resources for the Future, 1990. 182p. ISBN 0-915707-50-0.

This book is a collection of papers presented at a June 1988 workshop sponsored by Resources for the Future. Its purpose is to analyze the two major techniques for dealing with any future climate change: prevention and adaptation. The 14 chapters are organized into the three sections "Background," "Natural Resource Sectors," and "Perspectives." The book is a valuable source of some of the best thinking about these two approaches to the control of global warming.

Schneider, Stephen H. **The Coevolution of Climate and Life.** San Francisco: Sierra Club, 1984. 576p. ISBN 0-87156-349-5.

An earlier book on the topic than *Global Warming*, below, this title obviously lacks data obtained in the late 1980s. However, it provides a more complete discussion of the scientific and technical issues involved in global warming.

————. **The Genesis Strategy: Climate and Global Survival.** New York: Plenum Press, 1976. 419p. ISBN 0-306-30904-1.

Schneider provides a review of climatic changes through history, a discussion of the politics of climate research, and weather and climate modification. He points out how changing climate may affect growing conditions in the United States and result in a "North American grain drain." He suggests that one possible way of dealing with the "grain drain" is to develop programs for stockpiling grain during good growing seasons to provide needed supplies when growing seasons are less productive.

————. **Global Warming: Are We Entering the Greenhouse Century?** San Francisco: Sierra Club, 1989. 317p. ISBN 0-87156-693-1.

A very personal story about global warming written by one of the best known atmospheric scientists in the world today. The book is written in an informal, "chatty" style that is easy to read and to understand. It provides some fascinating insights into the day-to-day interactions that occur among scientists and politicians as they try to deal with this problem. An extensive collection of footnotes provides a very useful guide to additional publications in the field. If you had to choose only one book to read on this topic, this one would get my vote.

Seidel, Stephen, and Dale Keyes. **Can We Delay a Greenhouse Warming?** 2d ed. Washington, DC: Office of Policy Analysis, Office of Policy, Planning and Evaluation, 1983. 185p. No ISBN.

An early study by the U.S. government designed to find out "whether specific policies aimed at limiting the use of fossil fuels would prove effective in delaying temperature increases over the next 120 years." The study uses three models to estimate the effectiveness of various policy changes on the emission of greenhouse gases.

Silver, Cheryl Simon. **One Earth, One Future: Our Changing Global Environment.** Washington, DC: National Academy Press, 1990. 186p. ISBN 0-309-04141-4.

This book summarizes a 1989 conference on global environmental problems sponsored by the National Academy of Sciences. It is a highly readable summary of the status of a number of major environmental problems, climate change being one of them.

Silver, S. Fred, ed. **Global Climate Change: Human and Natural Influences.** New York: Paragon House, 1989. 424p. ISBN 0-89226-033-5.

This book is a collection of papers presented at a 1988 conference on global climate change. Papers deal with the effect of natural and anthropogenic carbon dioxide on climate, acid rain, nuclear winter, desertification, ocean pollution, and the Gaia hypothesis. Commentaries on many of the papers by other conference participants are also included.

Smith, J. B., and D. Tirpak, eds. **The Potential Effects of Climate Change on the United States,** 3 vols. New York: Hemisphere, 1990. 690p. ISBN 1-56032-071-0.

These three volumes appeared originally as a report by the Environmental Protection Agency, prepared in response to requests from the U.S. Congress. The initial study included original research, a review of existing studies, the identification of missing data, and a listing of policy options for dealing with climate change. The 18 chapters focus on regional effects in California, the Great Lakes, the Southeast, and the Great Plains and on possible impacts on water resources, sea-level rise, agriculture, forests, biological diversity, air quality, human health, urban infrastructure, and electricity demand. This is perhaps the most complete assessment of the effects of possible global climate change on the natural and human environment (at least in the United States) currently available.

Strain, Boyd R., and Jennifer D. Cure. **Direct Effects of Increasing Carbon Dioxide on Vegetation.** See Carbon Dioxide Research Division.

Tesar, Jenny E. **Global Warming.** New York: Facts on File, 1991. 111p. ISBN 0-8160-2490-1.

This text is intended for readers in grades 6 through 9. It provides a nicely balanced treatment of global warming, although the author seems convinced that climate change will be a serious problem in the future. Many useful suggestions for dealing with global warming are provided. The book is enhanced with drawings, charts, and black-and-white photographs.

Trabalka, John R. **Atmospheric Carbon Dioxide and the Global Carbon Cycle.** See Carbon Dioxide Research Division.

Trexler, Mark C. **Keeping It Green: Tropical Forestry and the Mitigation of Global Warming.** Washington, DC: World Resources Institute, 1992. 75p. ISBN 0-915825-65-1.

The author examines the relationship between tropical deforestation and global warming. He reviews a number of techniques, such as curbing deforestation, expanding plantations, promoting agroforestry, and restoring deforested lands, to help deal with problems of climate change.

————. **Minding the Carbon Store: Weighing U.S. Forestry Strategies To Slow Global Warming.** Washington, DC: World Resources Institute, 1991. 75p. ISBN 0-915825-48-1.

Tree planting often has been suggested as a way of helping to reduce global warming problems. The author reviews some of the specific alternatives available in the United States and assesses the potential benefit of each in affecting the greenhouse effect.

Union of Concerned Scientists. **Cool Energy: The Renewable Solution to Global Warming** and **The Global Warming Debate: Answers to Controversial Questions.** Cambridge, MA: Union of Concerned Scientists, n.d.

These two brochures attempt to present the fundamental facts about global warming in simple terms understandable to the nonscientist. A single copy of each is free.

U.S. Congress, Office of Technology Assessment. **Changing by Degrees: Steps To Reduce Greenhouse Gases.** Washington, DC: U.S. Government Printing Office, 1991. 354p. OTA-O-482.

A study prepared in response to the request of six congressional committees to answer the question, "Can the United States reduce carbon dioxide emissions in the near term?" The study makes a number of recommendations in the areas of energy supply; buildings, transportation, manufacturing, and forestry sectors; the food system; and international dimensions. Two useful appendixes outline a variety of scenarios based on recommendations presented in the report and a summary of state actions on greenhouse gas emissions.

Waggoner, P. E., ed. **Climate Change and U.S. Water Resources.** New York: John Wiley, 1990. 496p. ISBN 0-471-61838-1.

This is the third book in the Climate and Biosphere series, begun as a project of the Committee on Climate of the American Association for the Advancement of Science. The first section deals with background information on greenhouse gases and their projected effects on climate. The second and third sections focus on specific issues involving water availability, storage, transport, and use. Specific topics include the vulnerability of water systems to climate change and the effects of climate change on floods, droughts, irrigation, water quality, recreation and wildlife, urban water supplies, generation of electricity, and economics of water supplies and use.

Weiner, Jonathan. **The Next Hundred Years: Shaping the Fate of Our Living Earth.** New York: Bantam Books, 1990. 312p. ISBN 0-553-05744-8.

The author explains how the greenhouse effect occurs in the atmosphere. He then describes the methods scientists have used to study changes that take place in the atmosphere and how scientists learned that global warming may have begun. He explains how tropical deforestation may contribute to climate change. Finally, he suggests some possible future scenarios if climate does continue to change and recommends some actions people can take to reduce the threat of global warming. The book is especially appealing because it makes extensive use of personal anecdotes from the lives of scientists working with climate change research.

White, Margaret R. **Characterization of Information Requirements for Studies of CO$_2$ Effects: Water Resources, Agriculture, Fisheries, Forests and Human Health.** See Carbon Dioxide Research Division.

Young, Louise B. **Sowing the Wind: Reflections on the Earth's Atmosphere.** Englewood Cliffs, NJ: Prentice Hall, 1990. 207p. ISBN 0-13-083510-2.

As with many other books, this volume provides a general introduction to the causes of global warming, acid rain, and ozone depletion, with some suggestions for ways in which humans can deal with this problem in the future.

Reports

Various committees of the U.S. Senate and House of Representatives have held hearings on issues relating to climate change. The following is a list of some of the most important reports from these hearings.

Clean Air Act Amendments of 1987. Hearings before the U.S. Senate, Subcommittee on Environment and Public Works, Part 1, June 16–17, 1987. Washington, DC: U.S. Government Printing Office, 1987.

Federal Agency Response to Global Climate Change: The National Energy Policy Act of 1988. Hearings before the U.S. Senate, Committee on Energy and Natural Resources, August 11, 1988. Washington, DC: U.S. Government Printing Office, 1989.

Global Climate Changes: Greenhouse Effect. Hearing before the U.S. House of Representatives, Committee on Foreign Affairs, Subcommittee on Human Rights and International Organizations, March 10, 1988. Washington, DC: U.S. Government Printing Office, 1988.

The Global Environment Protection Act of 1988. Joint hearings before the U.S. Senate, Subcommittee on Hazardous Wastes and Toxic Substances and Committee on Environment and Public Works, Subcommittee on Environmental Protection, September 14 and 16, 1988. Washington, DC: U.S. Government Printing Office, 1988.

Global Environmental Change Research. Hearing before the U.S. Senate, Committee on Commerce, Science, and Transportation, Subcommittee on Science, Technology, and Space, July 16, 1987. Washington, DC: U.S. Government Printing Office, 1987.

Global Warming. Hearing before the U.S. Senate, Committee on Environment and Public Works, Subcommittee on Toxic Substances and Environmental Oversight, December 10, 1985. Washington, DC: U.S. Government Printing Office, 1986.

Global Warming. Hearing before the U.S. House of Representatives, Committee on Energy and Commerce, Subcommittee on Energy and Power, February 21, 1989. Washington, DC: U.S. Government Printing Office, 1989.

Greenhouse Effect and Global Climate Change. Hearings before the U.S. Senate, Committee on Energy and Natural Resources, November 9–10, 1987. Washington, DC: U.S. Government Printing Office, 1988.

Greenhouse Effect and Global Climate Change. Hearing before the U.S. Senate, Committee on Energy and Natural Resources, June 23, 1988. Washington, DC: U.S. Government Printing Office, 1988.

Modeling Greenhouse Climate Effects. Hearings before the U.S. Senate, Committee on Commerce, Science, and Transportation, Subcommittee on Science, Technology, and Space, May 8, 1989. Washington, DC: U.S. Government Printing Office, 1989.

National Energy Policy Act of 1988 and Global Warming. Hearings before the U.S. Senate, Committee on Energy and Natural Resources, August 11, September 19–20, 1988. Washington, DC: U.S. Government Printing Office, 1988.

National Global Change Research Act of 1989. Hearing before the U.S. Senate, Committee on Commerce, Science, and Transportation, February 22, 1989. Washington, DC: U.S. Government Printing Office, 1989.

Oversight Hearing on Implications of Global Warming for Natural Resources. Hearings before the U.S. House of Representatives,

Committee on Interior and Insular Affairs, Subcommittee on Water and Power Resources, September 27, 1988. Washington, DC: U.S. Government Printing Office, 1989.

Ozone Depletion, the Greenhouse Effect and Climate Change. Hearings before the U.S. Senate, Committee on Environment and Public Works, Subcommittee on Environmental Pollution, June 10–11, 1986. Washington, DC: U.S. Government Printing Office, 1986.

Ozone Depletion, the Greenhouse Effect and Climate Change. Joint hearings before the U.S. Senate, Committee on Environment and Public Works, Subcommittee on Environmental Protection and Hazardous Wastes and Toxic Substances, Part 2, January 28, 1987. Washington, DC: U.S. Government Printing Office, 1987.

The Potential Impact of Global Warming on Agriculture. Hearing before the U.S. Senate, Committee on Agriculture, Nutrition, and Forestry, December 1, 1988. Washington, DC: U.S. Government Printing Office, 1988.

In addition to the above hearing transcripts, a valuable source of information are the reports prepared by the Congressional Research Service when asked by House or Senate members to do so. Some of those reports are the following.

Congressional Research Service. **Agriculture, Forestry, and Global Climate Change—A Reader.** Prepared for the U.S. Senate, Committee on Agriculture, Nutrition, and Forestry. Washington, DC: Government Printing Office, April 1989.

———. **A Primer on Climatic Variation and Change.** Prepared for the U.S. House of Representatives, Committee on Science and Technology, Subcommittee on the Environment and the Atmosphere. Washington, DC: U.S. Government Printing Office, September 1976.

Congressional Research Service. **The Global Environment.** Susan R. Fletcher and Robert E. Morrison, eds. *CRS Review*, August 1989, pp. 1–26.

This journal contains background information on topics of concern to members of the U.S. Congress. This issue provides an extensive and useful background briefing on the global warming problem.

Gushee, David E. **Global Climate Change and the Greenhouse Effect: Congressional Activity and Options.** Congressional Research Service (CRS) Issue Brief, November 2, 1988.

Morrison, Robert E. **Global Climate Change.** CRS Issue Brief, October 6, 1989.

Tiemann, Mary, and Susan R. Fletcher. **International Environment: Overview of Major Issues.** CRS Issue Brief, October 13, 1989.

Newsletters

Greenhouse Effect Report (included in *World Environment Report*)
Business Publishers, Inc.
951 Pershing Drive
Silver Spring, MD 20910-4464
ISSN 0098-8235
Annual. $416 plus postage.

World Climate Change Report
Bureau of National Affairs
1231 25th Street, NW
Washington, DC 20037
ISSN 0957-9370
Discontinued. Past issues may still be available.

7

Selected Nonprint Resources

Films, Videocassettes, and Slide Programs

Climate Change Slides
Type: Slides
Age: All ages
Length: n/a
Cost: $1.00 per slide
Date: Various
Source: Information and Education Outreach Program
             National Center for Atmospheric Research
             P.O. Box 3000
             Boulder, CO 80307-3000

The National Center for Atmospheric Research has produced a number of colored slides illustrating various aspects of global warming research. Some examples of the topics for which slides are available include biomass burning, deforestation, methane research, tree rings (as a record of climate change), climate modeling, and predictions resulting from general circulation models.

Computer Simulation of the Global Climatic Effects of Increased Greenhouse Gases
Type: Videocassette
Age: High school and above
Length: 13 min.

Cost: Rental $20; purchase $50
Date: N.d.
Source: Information and Education Outreach Program
             National Center for Atmospheric Research
             P.O. Box 3000
             Boulder, CO 80307-3000

Through industrialization and deforestation, humans are increasing the global concentration of greenhouse gases such as carbon dioxide. If present trends continue, models predict that these activities will produce a noticeable change in the Earth's climate by the middle of the twenty-first century. This video shows how computer modeling is done and what some of its predictions are. The video may be rented from Modern Talking Pictures, Scheduling Center, 5000 Park Street North, St. Petersburg, FL 33709.

Future Conditional: Global Warming
Type: Videocassette
Age: High school and above
Length: 28 min.
Cost: Purchase $95
Date: 1992
Source: Documentaries
             Resources for the Future
             1616 P Street, NW
             Washington, DC 20036

A group of scientists and policy experts discuss questions such as what global warming is; how likely it is that climate change will occur in the twenty-first century; and what social, political, and economic consequences might accompany global change. The video attempts to present a balanced view about these topics and focuses especially on adapting to possible climate change and balancing economic growth with environmental concerns.

Global Warming: Hot Times Ahead?
Type: 16mm film, videocassette
Age: Grade 7 and up
Length: 23 min.
Cost: Rental $60; purchase $435 (film), $350 (video)
Date: 1990
Source: Churchill
             12210 Nebraska Avenue
             Los Angeles, CA 90025

This program describes the phenomenon of global warming and shows how the increase in greenhouse gases can be reduced in the short term by using energy more carefully and in the long term by developing new sources of energy that do not result in the release of carbon dioxide.

The Greenhouse Conspiracy

Type: Film
Age: Grade 7 and up
Length: 58 min.
Cost: Rental $50; purchase $350
Date: 1991
Source: Lucerne Films
37 Ground Pine Road
Morris Plains, NJ 07950

A very good presentation of the climate change issue with special attention to differences of opinion among scientists as to the degree to which global warming is actually occurring.

Greenhouse Crisis: The American Response

Type: Videocassette
Age: General audiences
Length: 11 min.
Cost: Purchase $14.95
Date: 1989
Source: Union of Concerned Scientists
26 Church Street
Cambridge, MA 02238

This award-winning video explains how problems of global warming have developed as a result of energy use in the United States. It outlines some of the steps individuals can take to reduce the problems of climate change in the future. The video is accompanied by a discussion guide.

The Greenhouse Effect

Type: Videocassette
Age: High school and above
Length: 50 min.
Cost: Rental $50; purchase $149
Date: 1988
Source: Films, Inc.
5547 Ravenswood Avenue
Chicago, IL 60640

Originally produced and broadcast by the BBC, this video explains how our extensive use of fossil fuels has produced the conditions that can lead to global warming. It then describes what effects might be expected of worldwide climate change. It outlines some possible scenarios for the year 2050 as well as discussing some methods for dealing with climate change.

The Greenhouse Effect

Type: 16mm film, videocassette
Age: High school and above
Length: 30 min.
Cost: Purchase $295
Date: 1989
Source: Landmark Films
              3450 Slade Run Drive
              Falls Church, VA 22042

This outstanding video thoroughly discusses every aspect of the climate change problem from an unbiased perspective. It describes the techniques used to study global warming and outlines some of the climate changes that have taken place in history. It reviews some of the predictions made for climate changes in the next century and reviews the steps that would have to be taken to alleviate those problems.

The Greenhouse Effect

Type: Slides, filmstrip
Age: Junior and senior high school
Length: 15 slides, 15 frames
Cost: Purchase $32.50 (slides, ES-38); $19.95 (filmstrip, ES-39)
Date: 1990
Source: William C. Brown
              2460 Kerper Boulevard
              Dubuque, IA 52001

The roles of human fossil-fuel use, deforestation, and air pollution are examined as sources of possible climate change in the future. The program considers how much global warming is likely to occur and how rapidly we can expect to see climate change beginning to take place.

Understanding the Greenhouse Effect and Global Climate Change

Type: Slides
Age: High school and above
Length: 53 slides
Cost: Rental free; purchase $25
Date: N.d.
Source: American Coal Foundation
              1130 17th Street, NW
              Suite 220
              Washington, DC 20036

This slide program is accompanied by a script, teacher's guide, and one-page summary. The program defines the greenhouse effect, discusses the role of each major greenhouse gas, describes what is meant by the theory of global climate change, discusses the tools used to assess and predict global climate change, and describes the factors that can influence climate change.

Computer Programs

Global Warming
Type: HyperCard stack
Age: Grades 4 through 12
Cost: $12
Date: 1990
Source: National Science Teachers Association
           1742 Connecticut Avenue, NW
           Washington, DC 20009
           Item #MS-90

The Global Warming HyperCard stack is designed for use with Apple Macintosh computers. It provides students with a method for learning about the problem of climate change by means of problem-solving exercises on the computer. The three issues addressed in the program are, "How might global warming affect your life?" "Why are scientists concerned with the problem?" and, "What can individuals do about the problem?" A supplementary guide is supplied to accompany the computer activity.

Hothouse Planet
Type: Interactive computer program
Age: Junior and senior high school
Cost: $70–$72 (depending on format)
Date: 1990
Source: William C. Brown
           2460 Kerper Boulevard
           Dubuque, IA 52001

This computer program allows students to control factors that contribute to the greenhouse effect and study the effects that occur when doing so. Users also can study the various predicted effects that can result from global warming. The program is based on predictions from NASA's Goddard Institute for Space Sciences.

Glossary

Some of the following terms are used in more than one branch of science with slightly different meanings. The definitions given here are those that apply to climate research. A list of acronyms and abbreviations follows the list of terms and their definitions.

ablation   The process by which snow and ice are lost from a glacier or ice field. Ablation occurs as the result of melting, evaporation, erosion, and the breaking off of large chunks of ice ("calving").

absorption coefficient   A measure of the amount of radiant energy that is absorbed per unit distance or per unit mass of a substance.

acclimation (also acclimatization)   The process by which an organism changes in order to adjust to a new environment.

adaptation   The process by which an organism or population adjusts to a new or changed environment. The process occurs when natural selection acts on genetic variation in the population.

adiabatic process   Any change in which there is no gain or loss of heat across the boundaries of a system.

advection   The flow of a gas (such as air) or a liquid (such as seawater) that causes a change in the temperature or some other physical property of the material.

aerosol   A form of matter that consists of tiny particles that range in size from about 10^{-3} to 10^{-2} micrometers in size.

afforestation   The process of planting a forest in an area where no forest originally existed.

airborne fraction   The proportion of a substance that remains in the atmosphere compared to the amount absorbed by the Earth's lithosphere and hydrosphere. For example, the airborne fraction of carbon dioxide is the portion of that gas produced on Earth that remains in the atmosphere, compared to the amount that is used by plants and dissolved in oceans and lakes.

albedo   The fraction of the light that strikes a surface and is reflected by it.

algae   Simple life forms that contain chlorophyll but lack other typical plant structures. Algae live primarily in the oceans and are the primary food for fish, whales, and small aquatic organisms.

alpine glaciers   Glaciers found in high mountain ranges.

alternative energy source   Any source of energy other than human, animal, or fossil-fuel power. Some examples of alternative energy sources include solar energy, wind power, geothermal power, and tidal power.

altithermal period   A geological period extending from about 8,000 to 4,000 years ago during which summers were warmer than they are at present and in which precipitation zones shifted toward the poles. The term also is used to describe other periods with similar climatic characteristics.

ammonia   A compound made of nitrogen and hydrogen in which the ratio of elements is one part nitrogen to three parts hydrogen (NH_3). The Earth's primitive atmosphere probably contained significant amounts of ammonia gas although the compound is essentially missing from the atmosphere today.

anaerobic bacteria   Bacteria that live in atmospheres lacking free oxygen. The bacteria obtain their energy by decomposing compounds that contain oxygen or from other types of compounds.

anomaly   Some factor or data that deviates from that which is expected.

anthropogenic   Produced as the result of human action. For example, large amounts of carbon dioxide are now being released into the atmosphere by such anthropogenic activities as the burning of fossil fuels, agriculture, and cement making.

appropriate technology   Any kind of technology that is inexpensive, simple, and easy for individuals to use. For example, one type of appropriate technology is a simple solar cooker that can be used by rural African families as a source of heat.

atmosphere   The envelope of gases that surrounds the Earth's surface. The term also is used as a standard unit of pressure measurement equal to the pressure exerted by a column of mercury 76.00 centimeters (29.92 inches) in height.

atmospheric general circulation models (AGCMs)   General circulation models that deal only with the atmosphere.

atmospheric window   That portion of the atmosphere's absorption spectrum between about 8 and 13 micrometers (3×10^{-4} and 5×10^{-4} inches) through which radiation can normally escape.

bathymetry   The science of measuring ocean depths, a process that often is carried out in order to determine the topography of the sea floor.

benthic region   The layer of the ocean nearest the bottom.

best estimate   The extrapolation from a climate model that appears to be the most likely to occur.

biogeochemical cycle   The chemical interactions that take place among the atmosphere, biosphere, hydrosphere, and lithosphere.

biological diversity   The variety of plant and animal life that exists on the Earth or in some particular region of the Earth. Scientists are interested in biological diversity because they believe that the greater diversity of organisms there are in an area, the healthier that region is likely to be.

biomass   The total mass of living organisms found within a particular area.

biosphere   That portion of the Earth that can support life.

biota   All of the living organisms found in a given area or at a given time.

B.P.   An abbreviation for the phrase "before the present," used to indicated periods in the Earth's history.

Btu   An abbreviation for "British thermal unit," a unit used in the measurement of heat energy.

"business as usual"   A situation in which individuals, industries, and societies continue to operate as they have in the past, with no changes made to deal with some existing or anticipated problem, such as global warming.

calorie   A unit used for the measurement of heat in the metric system. One calorie is the amount of heat gained or lost by one gram of water as its temperature changes by 1° Celsius. A second unit, the Calorie (with a capital C), is equal to 1,000 small calories.

carbohydrate   An organic compound made of carbon, hydrogen, and oxygen with the generalized formula $C_x(H_2O)_y$. Sugars, starches, and cellulose are among the most familiar types of carbohydrates.

carbon-12   The most abundant naturally occurring isotope of carbon.

carbon-13   An isotope of carbon that occurs naturally but in low abundance (about 1 percent of all carbon).

carbon-14 A naturally occurring radioactive isotope of carbon that is often used to measure the age of materials.

carbon-based resources All of those resources that can be used as fuels. Carbon-based resources include fossil fuels in all their forms and biomass.

carbon budget An accounting of the gain and loss of carbon between any two steps in the carbon cycle.

carbon cycle The complex series of reactions by which carbon passes through the Earth's atmosphere, biosphere, hydrosphere, and lithosphere. For example, plants remove carbon in the form of carbon dioxide from the atmosphere and use it to produce carbohydrates in living organisms. When those organisms die, the carbon is returned to the Earth as carbon dioxide, as fossil fuels (during decay), or as inorganic compounds such as calcium carbonate (limestone).

carbon density A measure of the amount of carbon per unit area for a given region of the Earth, as, for example, a particular ecosystem.

carbon dioxide (CO_2) A compound made of carbon and oxygen in the ratio of one part carbon and two parts oxygen, by weight.

carbon dioxide fertilization Increase in plant growth that results from an increase in the amount of carbon dioxide in the atmosphere, usually as a result of human activities.

carbon isotope ratio A measurement of the ratio of carbon-12 to carbon-13 or carbon-14. The carbon isotope ratio is used to determine the age of a material or changes that may have occurred in Earth history.

carbon monoxide (CO) A compound made of carbon and oxygen in the ratio of one part carbon and one part oxygen, by weight.

carbon pool The sum of all those resources that contain carbon as a major component.

carbon sink Any reservoir that takes up carbon released from some other part of the carbon cycle. For example, the atmosphere is a major carbon sink because much of the carbon dioxide produced elsewhere on the Earth ends up in the atmosphere.

carbon source Any part of the carbon cycle that releases carbon to some other part of the carbon cycle.

cellulose A complex organic carbohydrate that is the major constituent of plant materials such as grass, wood, cotton, and flax.

CFC An abbreviation for chlorofluorocarbon.

chlorofluorocarbon An organic compound containing carbon, chlorine, fluorine, and (usually) hydrogen. Chlorofluorocarbons are synthetically produced materials much in demand in industry because they tend to be nontoxic, nonflammable, noncorrosive, odorless, and chemically inert.

clear cutting   A forestry practice in which all of the trees in one area are cut at one time.

climate   The sum total of the weather conditions for a particular area over an extended period of time, at least a few decades.

climate change   Variations that take place in weather conditions over extended periods of time, usually at least a few decades.

climate lag   The delay that occurs in climate change as a result of some factor that changes only very slowly. For example, the effects of releasing more carbon dioxide into the atmosphere may not be known for some time because a large fraction of that carbon dioxide is dissolved in the ocean and only released to the atmosphere many years later.

climate model   Any model that attempts to simulate the behavior of the climate, allowing scientists to make predictions about future climatic conditions. Also see **general circulation models.**

climate sensitivity   The amount by which climate changes as a result of variations in a single factor such as temperature change or precipitation. The term also is used in climate modeling to indicate the amount of change that occurs in two different simulations as the result of changing a single factor in the model.

climate signal   A statistically significant difference between the control and the modified simulations in a climate model. Also see **signal-to-noise ratio.**

climate system   A term that refers to the five physical components responsible for the climate and climate changes. These five components are the atmosphere, biosphere, cryosphere, hydrosphere, and lithosphere.

climate variation   Changes that occur in one or more climatic variables over time.

climatic analog   A climatic situation that existed in the past that is similar to the present climate situation. Scientists use climatic analogs to make predictions about future climatic changes.

climatic optimum   The historical period extending from about 5000 to 2500 B.C. During this period, the average global temperature was warmer than it is now in nearly all parts of the world.

cloud   A mass of small droplets of water and/or ice crystals formed by the condensation of water vapor in the atmosphere.

cloud albedo   The amount of light reflected from a cloud. Cloud albedo depends on a number of factors, including size of ice crystals, amount of liquid water and ice within the cloud, thickness of the cloud, and the cloud's relative orientation to the sun.

cloud feedback   The relationship between cloudiness and surface air temperature. As the Earth's surface temperature increases, the rate of evaporation of water from the oceans and lakes increases. This increase, in turn, is followed by an increase in the number and density of clouds formed in the atmosphere. As more clouds form in the atmosphere, they tend to reflect more solar radiation back into space, reducing the Earth's temperature. Because this process does not always occur in such a straightforward manner, scientists are still unclear exactly how and to what extent cloud feedback affects the Earth's surface temperature.

cogeneration   The process by which two different and useful forms of energy are produced at the same time. For example, heat produced by the combustion of solid wastes can be sold commercially to homes, office buildings, and factories and also can be used to generate electricity.

combustion   The process of rapid oxidation, or "burning."

conservation   The wise use of natural resources so significant amounts of those resources will be available for future generations.

convection   The movement of heat from one place to another as the result of the transfer of a heated mass of liquid or gas. In climate studies, convection usually refers to the vertical transport of gases and/or liquids in the atmosphere and/or the oceans.

Coriolis effect   The tendency for an object in motion above the Earth to turn to the right in the Northern Hemisphere and to the left in the Southern Hemisphere. The effect is caused by the Earth's rotation on its axis.

country study   Any official national study on greenhouse gas emissions, impact assessments, and emission mitigation analysis.

coupled general circulation models (CGCMs)   Advanced forms of general circulation models that attempt to analyze the interaction of atmosphere and ocean.

cryosphere   That portion of the Earth's surface that is occupied by masses of ice and snow. Included in the cryosphere are the continental ice sheets; mountain glaciers; sea, lake, and river ice; and snow cover on land.

cultivar   Any type of plant that can be produced by cultivation.

decay   The oxidation of dead organic matter. Two important products of decay are carbon dioxide and methane.

decomposers   Plants and animals that survive by extracting energy from dead protoplasm. The vast majority of decomposers are fungi and bacteria.

decomposition   The process by which organisms (usually bacteria and fungi) break down dead organic matter into simpler compounds.

deep water   The technical term for that part of the ocean that lies below the main thermocline.

deforestation   The removal of forests from the land. Deforestation occurs when humans cut down trees to use as fuel or for building materials or to obtain land for agricultural, developmental, or other purposes. The loss of trees through deforestation results in the loss of an important sink for carbon dioxide, although the overall climatic effects of the practice are not entirely understood.

dendrochronology   The practice of determining the time of occurrence of past events by studying the growth rings in trees. The spacing and width of rings provides information about climatic conditions present at the time the tree was growing.

desertification   The process by which a once productive area of land is converted into a desert as a result of human activities. For example, overgrazing on rangelands, often accompanied by an extended period of drought, may convert those rangelands into desert.

developed nation   A term sometimes used to describe a nation that tends to have (1) a relatively low birth rate, (2) a relatively high per capita income, (3) a relatively low fraction of its people engaged in agriculture, and (4) a relatively strong economy.

developing nation   A term sometimes used to describe a nation that tends to have (1) a relatively high birth rate, (2) a relatively low per capita income, (3) a relatively high fraction of its people engaged in agriculture, and (4) a relatively weak economy. The term *less developed country* (LDC) is sometimes used to compare such nations with more developed countries (MDCs).

downwelling   The process by which warm, nutrient-filled surface waters collect along a coastline and sink downward into cooler waters below.

eccentricity   The flattening of the Earth's orbit around the sun. The Earth's eccentricity increases and decreases in a cycle that takes about 100,000 years, accounting for varying amounts of solar radiation reaching the Earth's surface over time.

ecosystem   A system of plants and animals that interact with each other and with their physical surroundings.

El Chicon   An active volcano in Mexico that last erupted in 1983.

El Niño   A change that occurs in the flow of ocean currents off the western coast of South America. The name for the phenomenon comes from the fact that the change tends to occur around Christmastime (El Niño is Spanish for "Christ child"). El Niño may bring about significant changes in ocean temperatures and weather patterns extending as far north as the west coast of North America.

emission mitigation analysis   Any study that attempts to predict the effects of reducing the release of one or more greenhouse gases into the atmosphere.

emissions   Materials released as waste products from human activities. Emissions include gases, particulates, vapors, solids, and other materials released from smokestacks, chimneys, automobile exhaust systems, and similar sources.

emissivity   The ratio of the amount of radiation emitted by a surface compared to the amount emitted by a black body at the same temperature.

endangered species   Any species having so few individual members that its survival is seriously threatened.

environment   The sum total of all the external conditions that affect the life of a population or an organism.

equilibrium   A state in which two opposing forces exactly balance each other.

equivalent carbon dioxide   The combined effect of greenhouse gases other than carbon dioxide, including methane, nitrogen oxides, and CFCs. The behavior of these gases in the environment often is poorly understood, so for the purposes of simplicity, they are lumped together and considered to have an effect equivalent to that of an equal concentration of carbon dioxide.

euphotic zone   That layer of an ocean or some other body of water that receives enough sunlight to permit photosynthesis to occur. The depth of the euphotic zone varies depending on factors such as the cloudiness of the water and the sun's angle, but it tends to be about 80 meters (260 feet).

evaporation   The process by which a liquid changes to a vapor.

evapotranspiration   The process by which water is released from the Earth's surface by evaporation from bodies of water and by transpiration from plants.

extrapolation   A prediction about some future behavior based on what is known about past and present trends. For example, scientists often try to extrapolate the likely concentration of atmospheric carbon dioxide in the twenty-first century, based on changes that have occurred in the past.

feedback mechanism A series of changes in which the final step results in an outcome that influences the first step in the series. See also **positive feedback** and **negative feedback.**

first detection The initial recognition that a significant change has occurred in a factor involved in climatic change. An investigator must be able to determine that the change is not a result of normal random variations in the climate. Also see **signal-to-noise ratio.**

forcing factor Some factor that, when it changes, can bring about a change in the climate. For example, temperature is a forcing factor because increasing or decreasing temperatures will have some effect on future climate.

fossil fuel Any naturally occurring mixture of pure carbon and hydrocarbons that can be used as a fuel. Coal, petroleum, and natural gas are the most common forms of fossil fuels.

Gaia The Greek mother-goddess. The term has been used by the British scientist James Lovelock to describe his theory that human life exerts an important influence on the planet, changing land, air, and sea to make the planet more hospitable to human life. The Gaia theory has been the subject of intense scientific debate in the past two decades.

gamma rays A form of electromagnetic radiation with very short wavelength and very high frequency. Gamma rays are a component of solar radiation.

general circulation models (GCMs) Models (usually computer programs) that use fundamental laws of physics and chemistry to analyze the interaction of temperature, pressure, solar radiation, and other climatic factors in order to predict climatic conditions for the past, present, or future.

geosphere Another name for the lithosphere.

geothermal power Energy that is obtained by making use of heat stored in rocks beneath the Earth's surface. The heat may be tapped directly or it may be obtained from hot water and/or steam formed in the rocks.

glacial maximum The greatest advance of a glacier, measured in either distance or time.

glaciation The process by which a large area of land or water is covered with ice and snow for a significant period of time.

glacier A large mass of ice and snow that forms on land or water and that survives for a relatively long period of time. From time to time in Earth history, very large glaciers have covered significant portions of the continents for thousands of years. Today most glaciers are found in limited areas in high mountainous regions.

global warming The process by which the Earth's annual average temperature increases by a significant amount over a relatively extended period of time. In today's world, the threat of global warming appears to result from the release of carbon dioxide, methane, and other greenhouse gases as a result of human activities.

global warming potential (GWP) A concept developed by scientists to measure the possible warming effect on the climate of various greenhouse gases.

greenhouse decade A term sometimes used for the 1980s, a decade in which several of the warmest years in recorded history were observed.

greenhouse effect A term used to describe the effect on the Earth's temperature that results from the capture of heat by molecules of carbon dioxide, water vapor, and other gases in the Earth's atmosphere. Scientists believe that, without the greenhouse effect, the Earth's surface temperature would be about 30° C (54° F) less than it actually is.

greenhouse gas Any gas that does not absorb solar radiation but does absorb long-wavelength radiation reflected from the Earth's surface. The most important greenhouse gases are water vapor, carbon dioxide, nitrous oxide, and methane.

GtC The abbreviation for gigatons of carbon, equivalent to 1 billion tons of carbon.

gyre A circular pattern of water flow in the open seas.

halocarbons Hydrocarbons that contain one or more halogen (usually fluorine or chlorine) atoms.

heat balance A mathematical accounting for all the heat that enters and leaves the Earth's atmosphere.

heat flux The amount of heat transferred across a unit area of surface in a unit of time.

heat island A dome-shaped region of the atmosphere over an urban area where the temperature is higher than surrounding areas. The elevated temperature is caused by heat absorbed by cement, asphalt, concrete, and other materials used in buildings, pavement, highways, and other human-made structures.

heat sink A reservoir in which heat is stored for an extended period of time. The oceans are one important heat sink on the Earth.

hectare A unit of area measure in the metric system, equal to 10,000 square meters (108,000 square feet).

hydrocarbon A compound containing only hydrogen and carbon. Hydrocarbons are major components of all fossil fuels.

hydrologic budget A quantitative accounting of all the water that flows into, flows out of, and is stored within a given basin or area over a given time period.

hydrologic cycle A series of changes involving water in the atmosphere, hydrosphere, and lithosphere. Some of the steps in the hydrologic cycle include water evaporation from oceans, lakes, and rivers; the water condensation in the form of clouds; and water precipitation in the form of rain, snow, hail, and sleet.

hydrology The science that deals with the properties, distribution, and circulation of water.

hydrosphere All that portion of the Earth that consists of water. The hydrosphere includes but is not limited to the oceans; freshwater lakes, rivers, streams, ponds, and groundwater; and water vapor in the atmosphere.

hypsithermal period The name given to a period about 4,000 to 8,000 years ago when the Earth's average annual temperature was higher than it is today. See also **altithermal period.**

ice age Any period in Earth history when significant portions of the Earth's surface were covered by glaciers.

ice albedo The amount of light reflected by ice.

ice core A cylindrical section of ice removed from a glacier or an ice sheet in order to study climate patterns of the past.

ice cover The thickness of glacial ice covering a specific land area.

ice sheet The name given to any glacier that covers more than 50,000 square kilometers (19,300 square miles) to a significant depth.

ice shelf A thick sheet of ice that is attached to the land on one side but that floats freely on the opposite side.

impact assessment An attempt to predict how changing one or more factors that have an influence on an area or situation would affect that area or situation.

incursion (of seawater)   The flow of seawater into a freshwater table as a result of the draining of the water table or a rise in the level of seawater.

Industrial Revolution   The change from an agricultural to an industrial society and from home-based to factory-based manufacturing that took place in England between about 1750 and 1850.

industrial smog   A form of smog that results from the combination of smoke and fog.

infrared radiation   Electromagnetic radiation in the range between about 0.7 micrometer and 1,000 micrometers (3×10^{-5} and 4×10^{-2} inches).

insolation   The amount of solar radiation that falls on a unit area of horizontal surface in a given period of time.

interglacial period   The period of time between two glacial periods. Temperatures during an interglacial period tend to be significantly warmer than they are during a glacial period.

interpolation   The process of estimating missing data that exists between two or more known pieces of data.

irradiance   The total amount of radiant power that falls on a unit of area. Also called the radiant flux density.

isotherm   A line on a map that connects all points of equal temperature.

isotopes   Two or more atoms that have the same atomic number (the same number of protons) but different atomic masses (different numbers of neutrons).

lapse rate   In general, the rate at which any meteorological condition changes with elevation. The term is most commonly used to refer to temperature changes with elevation. The normal lapse rate is 3.6° C (6.5° F) per 1,000 feet change in elevation.

latent heat   The amount of heat absorbed or released by a substance during any change of state (melting, freezing, vaporization, etc.). The term is also used to refer to the amount of energy transferred to and from the Earth's surface to the atmosphere as a result of evaporation and condensation.

Le Chatelier's principle   A scientific law that states that a system in equilibrium adjusts in such a way as to minimize the effect of any stress applied to it.

less developed country (LDC)   A term sometimes used to describe a nation with low per capita income, a high rate of population growth, a weak economy, a large fraction of its population involved in agricultural activities, and a high rate of adult illiteracy.

lithosphere The solid part of the Earth, consisting of rocks, soil, and sediments.

Little Ice Age A period in Earth history that lasted from about 1550 to about 1850 in the Northern Hemisphere, when temperatures were somewhat colder than usual and glaciers extended over relatively large areas of North America, Asia, and Europe.

longwave radiation Electromagnetic radiation having wavelengths greater than 4 micrometers (1.6×10^{-4} inches). Because this type of radiation is the form emitted from the Earth, it is sometimes referred to as terrestrial radiation.

Mauna Loa A volcano on the island of Hawaii that last erupted in 1984. The Mauna Loa Observatory has maintained the longest continuous collection of reliable daily atmosphere records (carbon dioxide concentrations) in the world.

mean sea level The average height of the seas' surface as measured over a 19-year period.

mesosphere The layer of the atmosphere that extends from an altitude of 50 kilometers to about 85 kilometers (30 miles to about 53 miles) above the Earth's surface.

metabolism The sum total of all chemical reactions required to keep an organism alive.

methane (CH_4) A compound of carbon and hydrogen produced by the decomposition of organic matter. Methane is the largest single component of natural gas.

Milankovitch theory An astronomical theory developed by the Serbian mathematician Milutin Milankovitch that shows how climate change is affected by seasonal and geographical changes in the amount of solar energy reaching the Earth's surface.

"missing sink" The sink for carbon dioxide that has not yet been found that can account for the discrepancy between the amount of carbon dioxide produced on Earth and that which can be accounted for in known sinks.

model A mathematical or physical representation of some object, situation, or process. Scientists use models to study phenomena that are too large, too small, too complex, or too difficult to study directly for some other reason.

more developed country (MDC) A term sometimes used to describe a nation with high per capita income, a low rate of population growth, a strong economy, a small fraction of its population involved in agricultural activities, and a low rate of adult illiteracy.

Mount Pinatubo   A volcano in the Philippine Islands that erupted in 1991. Scientists study volcanic eruptions such as this one to determine the effectiveness of their climate models. By one estimate, the eruption of Mount Pinatubo has significantly changed (cooled) the Earth's climate to such an extent that other long-term climate changes may be undetectable for about five years.

nanometer (nm)   One-billionth (10^{-9}) of a meter.

negative feedback   A series of reactions in which the final step acts in such a way as to reduce or dampen the first step in the sequence.

nitrous oxide (N_2O)   A compound of nitrogen and oxygen produced during the combustion of fossil fuels, by the breakdown of chemical fertilizers, and by bacterial action, especially in tropical soils.

nuclear energy   A form of energy produced by fission reactions in isotopes of uranium or plutonium.

nutrient   Any substance that is required by an organism for its survival and growth.

ocean currents   The movement of large masses of surface water in the oceans, caused largely by winds blowing over the surface water.

ocean general circulation models (OGCMs)   General circulation models that deal with the temperature and other changes in the ocean.

ocean mixing   Any process that involves the mixing of ocean waters.

opacity   The degree to which light is blocked out.

outgasing   The release of a gas from some solid or liquid, such as the escape of carbon dioxide from the surface of seawater.

oxidation   A chemical reaction in which oxygen reacts with some substance. Combustion (burning), decay, and rust are common examples of oxidation.

oxides of nitrogen   A term used to describe all possible oxides of nitrogen (there are five). The term is often abbreviated as NO_x.

ozone   An allotrope (form) of oxygen that consists of three atoms to the molecule (O_3) rather than the more common two atoms to the molecule (O_2). Ozone is a normal component of the atmosphere and forms a layer in the stratosphere that absorbs ultraviolet radiation from sunlight. It is also produced during the combustion of fossil fuels on the Earth's surface. One of the primary mechanisms by which it forms involves the interaction of sunlight with oxides of nitrogen released from the exhaust systems of internal combustion vehicles.

ozone hole A region in the stratosphere over the Antarctic where the density of ozone has suddenly decreased during the spring over the past decade or so.

palaeoanalog A similarity between some event or situation that occurred in the past and one that exists presently or that is predicted by a climate model.

palynology The study of plant spores and pollen, especially those in the fossil state. This study makes it possible for scientists to reconstruct earlier climates based on the kinds of plants that lived during earlier periods of Earth history.

parametrization The use of a model to represent some physical system with factors that are admittedly oversimplified in order to better understand that system.

particulate matter Very small pieces of solid or liquid matter, such as soot, dust, fumes, or mist.

parts per billion (ppb) A unit of concentration measure that can be expressed for volume (ppbv) or weight (ppbw).

parts per million (ppm) A unit of concentration measure that can be expressed for volume (ppmv) or weight (ppmw).

past climate analogs The reconstruction of past climates based on comparisons with modern climates that occur at different elevations or longitudes.

pCO$_2$ The partial pressure of carbon dioxide. Partial pressure is the pressure that would be exerted by a gas if it existed by itself in a container.

perturbation Some change that occurs in a system because of the presence of a particular substance or event. For example, the increase in carbon dioxide in the atmosphere is a major cause of perturbation in the Earth's climate.

photochemical smog A form of smog produced when hydrocarbons and oxides of nitrogen are converted to noxious chemicals by sunlight.

photosynthesis The chemical reaction in which plants manufacture carbohydrates from carbon dioxide and water, using chlorophyll as a catalyst. Oxygen gas is produced as a by-product of the reaction.

photovoltaic power The production of an electrical current by some substance as a result of its being exposed to light or some other form of electromagnetic radiation.

phytoplankton That part of the plankton community that consists of tiny plants, primarily algae and diatoms.

planetary albedo The fraction of the total amount of solar radiation reaching a planet that is reflected back into space. That fraction is about 30 percent for the Earth.

plankton Small organisms that float in water, especially on its surface.

polar ice caps The extensive regions of deep ice and snow that cover the areas surrounding the North and South Poles.

positive feedback Any series of reactions in which the final step of the series produces some change that amplifies the first step of the series.

precession The tendency of the Earth's axis to wobble in space over a 23,000-year period. The Earth's precession is one of the factors that results in the planet receiving different amounts of solar energy over very long periods of time.

precipitation The depositing of any form of solid or liquid from the atmosphere to the Earth's surface. Some typical forms of precipitation include rain, snow, hail, and sleet.

precursor Some chemical substance from which another substance is made. For example, it appears that ozone is destroyed in the stratosphere when it is attacked by atomic chlorine. The atomic chlorine is produced when CFCs are decomposed by sunlight. CFCs, therefore, are a precursor of atomic chlorine in the atmosphere.

primitive atmosphere A term used to describe the Earth's atmosphere during its earliest history. That atmosphere is thought to have consisted largely of reducing gases such as hydrogen, ammonia, methane, and carbon monoxide.

protocol An early draft from which an agreement, such as a treaty, is developed. Also, an outline or plan that describes the procedures to be followed in some kind of study or research.

proxy climate indicators Any type of data that provides information about climatic conditions during a specific time period. Some examples of proxy climate indicators include tree rings, fossilized remains, and isotope concentrations.

pyrgeometer An instrument used to measure the amount of radiation reaching the Earth's surface from outer space.

radiant flux density See **irradiance.**

radiation balance The difference between the amount of radiation absorbed by the Earth's surface and the amount re-emitted as infrared radiation. Also known as the radiation budget.

radiatively active gas Any gas in the atmosphere that absorbs incoming solar radiation or outgoing infrared radiation reflected from the Earth's surface.

radiosonde A balloon-borne instrument used to collect and transmit meteorological data in the upper atmosphere. Radiosondes operate to a maximum altitude of about 30 kilometers (19 miles).

random turbulence The apparently disorganized movement of a fluid (gas or liquid). The tumbling of water in a stream is an example of random turbulence.

reflectivity The ratio of the energy carried by a wave that is reflected from a surface to the energy of a wave that falls on the surface.

reservoir A place where anything is stored. A lake, for example, is a natural reservoir for fresh water.

residence time The time during which some substance remains in a particular reservoir. Residence time also can be expressed in quantitative terms as the total mass of the reservoir divided by the total flux of mass into or out of that reservoir.

respiration A series of chemical reactions by which organisms use oxygen from the environment to oxidize food, producing carbon dioxide as a waste product.

rocketsonde A rocket-borne instrument used to collect and transmit meteorological data in the upper atmosphere. Rocketsondes operate to a maximum altitude of about 76 kilometers (47 miles).

Sahel region A semiarid area in Africa immediately south of the Sahara Desert.

salinity The total amount of dissolved solids in seawater, usually expressed in parts per thousand by weight in 1 kilogram of seawater.

scenario A plan, design, or prediction regarding some future outcome. For example, scientists can imagine various scenarios for future climate change depending on a variety of changes in factors such as temperature or amount of solar radiation.

sea surface temperature The temperature of the upper 0.5 meter (1.6 feet) of seawater.

seasonal variation The changes that occur in meteorological conditions averaged over a period of three months.

secular carbon dioxide trend The steadily increasing concentration of carbon dioxide in the atmosphere. This trend has been best confirmed by measurements made at the Mauna Loa Observatory in Hawaii.

self-correcting mechanism   Any process that operates in such a way as to reduce or eliminate some trend in the process itself. For example, increasing concentrations of carbon dioxide in the atmosphere may result in a warmer global temperature. But an increase in the global temperature may also result in increased evaporation of ocean waters and increased cloud formation. Increased cloud volumes may then cause more solar radiation to be reflected back into space, reducing the Earth's annual average temperature. The warming process induced by higher levels of carbon dioxide may, therefore, have a self-correcting factor built into it.

sensible heat   In meteorology, the amount of radiative energy that has been given up by the Earth's surface to the atmosphere. The term has other specialized meanings in other fields of science.

sequestration   The process by which a substance is removed from the free state and tied up in some other material. For example, carbon dioxide is sequestered—removed from the atmosphere—when it is used by green plants to make carbohydrates during the process of photosynthesis.

shortwave radiation   Solar radiation with wavelengths of less than 4 m. Also known as solar radiation.

signal-to-noise ratio   A quantitative measurement of the significance of changes in any one climatic factor compared to its normal, random level. For example, scientists can measure the concentration of carbon dioxide in the Earth's atmosphere over some extended period of time. That measurement provides information about the background level, or noise, for carbon dioxide. Measurements that significantly differ from this background level, then, stand out as a signal of some change that is taking place.

silviculture   The science dealing with the management of forest resources.

simulation   An attempt to imitate or reproduce some physical phenomenon using mathematical or physical models or both. For example, general circulation models attempt to simulate the Earth's climate now or in the past, as an attempt to predict future climates.

sink   The final location in which some material is deposited for an extended period of time. For example, the oceans are an important sink for carbon dioxide because the gas dissolves in water and tends to remain dissolved for long periods of time.

smog   A form of air pollution that gets its name from its resemblance to *smoky fog*. Two forms of smog exist: industrial smog and photochemical smog.

snow albedo The fraction of sunlight that is reflected by snow.

solar constant The rate at which solar energy strikes the outermost layer of the Earth's atmosphere. Technically, the solar constant is defined as the rate at which solar energy is received perpendicularly on one square centimeter of the atmosphere's surface when the Earth is at its average distance from the sun. The value of the solar constant is 0.140 watt/cm^2.

solar cycle The periodic change in the number of sunspots. The interval between successive minima or maxima is about 11.1 years.

solar radiation See **shortwave radiation.**

starch A type of carbohydrate formed in green plants during photosynthesis.

stratosphere That region of the atmosphere between the troposphere and the mesosphere. It extends from an altitude of about 15 to 50 kilometers (9 to 30 miles) above the Earth's surface.

sulfate aerosols Particulate matter that consists of compounds of sulfur formed by the interaction of sulfur dioxide (SO_2) and sulfur trioxide (SO_3) with other compounds in the atmosphere.

sunspot A region on the sun's surface that appears to be darker, and therefore cooler, than the surrounding area.

sunspot cycle See **solar cycle.**

surface air temperature The temperature of air near the surface of the Earth. In practice, the measurement is usually made at a distance of about 2 meters (6.5 feet) above the Earth's surface.

surface albedo Of the total solar radiation that strikes the Earth's surface, the fraction that is reflected by the Earth's surface.

technological fix An attempt to solve a problem by some technical means, rather than by instituting social, economic, political, or other types of changes. For example, one method for reducing carbon dioxide concentrations in the atmosphere is to place a tax on the burning of fossil fuels. This is an economic approach to the problem. Another method is to spread iron powder on the oceans' surface, to increase the metabolism of plankton and thus remove more carbon dioxide from the atmosphere. This represents a technological fix for the problem.

terrestrial radiation The total amount of radiation emitted by the Earth, including its atmosphere, in the temperature range of about $-70°$ C to about $30°$ C (about $-90°$ F to about $90°$ F).

thermocline A zone of ocean water where the temperature decreases more rapidly than it does above or below that zone. The permanent thermocline separates the warm surface layer of the ocean from the cold deep layer. It begins anywhere from 10 to 500 meters (30 to 1500 feet) below the surface and may extend to 1500 meters (5000 feet) or more in depth.

thermosphere The uppermost region of the atmosphere, beginning at an altitude of about 80 kilometers (50 miles) above the Earth's surface.

tilt The orientation of the Earth's axis in space. The Earth's tilt changes from a maximum of 24.5° to a minimum of 21.5° over a 41,000-year period. This change in tilt is one factor that results in the planet receiving different amounts of solar radiation at different periods.

TIROS The name given to a group of satellites whose primary function is to observe weather patterns on the planet.

trace gas Any one of the less common gases found in the Earth's atmosphere. Nitrogen, oxygen, and argon make up more than 99 percent of the Earth's atmosphere. Other gases, such as carbon dioxide, water vapor, methane, oxides of nitrogen, ozone, and ammonia, are considered trace gases. Although relatively unimportant in terms of their absolute volume, they have significant effects on the Earth's weather and climate.

transient model A model in which climate changes as a result of increasing greenhouse gases are tracked over time.

transpiration The process by which plants give off water from their leaves.

tree ring The amount by which a tree grows in circumference in a single year. Tree rings provide important data about weather and climate because tree growth varies as a consequence of changes in precipitation, temperature, solar radiation, and other factors. The science of estimating past climates based on tree ring studies is known as dendrochronology or dendroclimatology.

trophic level A segment of a food chain or food web in which all organisms obtain food and energy in essentially the same way. For example, all the organisms in an area that obtain their food from plants (that is, the herbivores) are at the same trophic level.

tropical rain forest Any forested region in the tropical zone. Also known as a selva.

tropopause The region between the troposphere and the stratosphere in the atmosphere, at an altitude of about 15 kilometers (9 miles) above the equator and about 8 kilometers (5 miles) above the poles.

troposphere The lowest layer of the atmosphere, extending from the Earth's surface to an altitude of about 15 kilometers (9 miles).

ultraviolet radiation Electromagnetic radiation with a wavelength of about 4 to 400 nanometers.

upwelling The vertical movement of ocean water that brings cold subsurface water to the surface. The climatic importance of upwelling is that it brings carbon dioxide dissolved in cold, deep water to the ocean's surface, where the carbon dioxide then is released to the atmosphere.

urbanization The movement of people, usually from rural areas, to regions of more concentrated population such as cities.

vapor The gaseous form of any material that is normally a solid or liquid.

visible light A form of electromagnetic radiation with a wavelength of about 370 to 730 nanometers.

water vapor Water in its gaseous form.

water vapor feedback A process by which an increase in the amount of water vapor in the atmosphere tends to increase the atmosphere's absorption of longwave radiation, thereby contributing to a heating of the atmosphere.

weather The state of the atmosphere as defined primarily by six factors: temperature, barometric pressure, wind velocity and direction, humidity, clouds, and precipitation.

weather satellite A satellite used primarily for the observation of weather on the Earth.

wetlands Any area that is continuously saturated with water for some period of time each year.

wind power An alternative energy source made possible by harnessing the energy of moving air using various types of windmills.

Xrays A form of electromagnetic radiation with wavelengths of about 1 to 10 nanometers.

zooplankton That part of the plankton community consisting of tiny animals.

Acronyms and Abbreviations

Acronyms and abbreviations are widely used to describe organizations, programs, chemicals, and other things having to do with global warming. Listed below are a number of those acronyms and abbreviations. The names of some organizations and programs are followed by the abbreviation for their parent agency or their national or international affiliation in parentheses.

AAAS   American Association for the Advancement of Science

BEPS   building energy performance standards

Btu   British thermal unit

CAFE   corporate average fuel efficiency

CEES   Committee on Earth and Environmental Sciences (U.S. Office of Science and Technology Policy)

CEOS   Committee on Earth Observations Satellites

CFCs   chlorofluorocarbons

CH$_4$ methane

CISC   Committee on the Impacts of Stratospheric Change (NAS)

CO$_2$ carbon dioxide

CRU   Climate Research Unit, University of East Anglia, Norwich, U. K.

DOA   See USDA

DOE   Department of Energy (U.S.)

EEC   European Economic Community

EIA   Energy Information Administration (DOE)

ENSO   El Niño–Southern Oscillation

EOS   Earth Observing System (international satellite observing system to be in operation by 1998)

EPA   Environmental Protection Agency (U.S.)

EPRI   Electric Power Research Institute

FAO   Food and Agriculture Organization (UN)

FDA   Food and Drug Administration (U.S.)

FIRE   First International Satellite Cloud Climatology Project (ISCCP) Regional Experiment

GCM general circulation model

GEWEX Global Energy and Water Cycles Experiment

GFDL Geophysical Fluid Dynamics Laboratory (NOAA)

GISS Goddard Institute for Space Studies (NASA)

GNP gross national product

GTCE Global Tropospheric Chemistry Experiment (IGBP)

HCFC hydrochlorofluorocarbon

ICSU International Council of Scientific Unions

IGAC International Global Atmospheric Chemistry Programme (IGBP)

IGBP International Geosphere-Biosphere Programme

IPCC Intergovernmental Panel on Climate Change (UN)

JGOFS Joint Global Ocean Flux Study

MIT Massachusetts Institute of Technology

NAS National Academy of Sciences (U.S.)

NASA National Aeronautics and Space Administration (U.S.)

NCAR National Center for Atmospheric Research (NOAA)

NGO nongovernmental organization

NOAA National Oceanic and Atmospheric Administration (U.S.)

NO_x nitrogen oxides

NRC National Research Council (NSF)

NSF National Science Foundation (U.S.)

O_3 ozone

OMB Office of Management and Budget (U.S.)

OTA Office of Technology Assessment (U.S. Congress)

PAGES Past Global Changes

SCOPE Scientific Committee on Problems of the Environment

SERI Solar Energy Research Institute (U.S.)

SIO Scripps Institute of Oceanography (NOAA)

TOGA Tropical Ocean–Global Atmosphere Programme (WCRP)

UN United Nations

UNCED United Nations Conference on Environment and Development ("Earth Summit")

UNEP United Nations Environment Programme

USDA U.S. Department of Agriculture

USGCRP U.S. Global Change Research Program

WCRP World Climate Research Programme, World Meteorological Organization

WMO World Meteorological Organization (UN)

WOCE World Ocean Circulation Experiment (WCRP)

Index

169

David Newton is a free-lance writer with more than 450 commercial publications to his credit, including about 50 books. His most recent titles include *Gun Control, James Watson and Francis Crick: Discovery of the Double Helix and Beyond, Population: Too Many People?, AIDS Issues: A Handbook,* and *Cities at War: Tokyo.* Dr. Newton formerly taught junior and senior high school science and mathematics in Grand Rapids, Michigan, and chemistry, teacher education courses, and human sexuality at Salem (Massachusetts) State College. He is currently Adjunct Professor in the College of Professional Studies at the University of San Francisco. In his spare time, he is a volunteer for the San Francisco SPCA and for Open Hand, an organization that provides meals for HIV positive men and women.